Understanding & Using the Oscilloscope

by

Clayton Hallmark

SECOND EDITION

FIRST PRINTING—JULY 1973

Printed in the United States
of America

Hardbound Edition: International Standard Book No. 0-8306-3664-1

Paperbound Edition: International Standard Book No. 0-8306-2664-6

Library of Congress Card Number: 73-84546

Preface

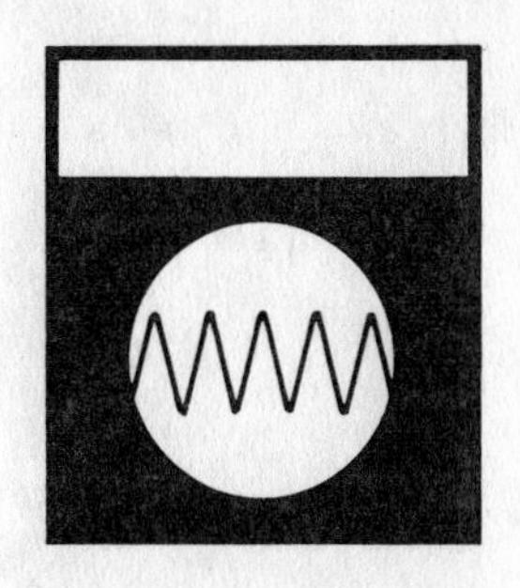

This book was written for the electronics technician who has an intermediate knowledge of troubleshooting and servicing fairly complex circuitry. We envisioned a person who uses a scope in his work but who blindly follows the scope setup instructions supplied by the manufacturer. Assuming you want and need to know much more, this book takes you behind the scenes, so to speak, and tells you **what is happening** when you turn a knob on a scope's panel, **why** you need to do certain things to obtain a proper and **valid** display, **how** poor setup resulting in a poor display can lead you down byways rather than directly to sources of circuit trouble—in other words, it is a short but intensive course in knowing what's inside your scope, how to use it properly, and how to figure out what you're looking at after you get it.

Naturally, we describe various modes, circuits to be tested, input and output results vs expectations, and so on. All this is presented in step-by-step fashion—usually with simple algebra—and every section builds logically on those preceding it.

With the knowledge you will gain (or perhaps, deeper knowledge, if you are reviewing), you will be ready to understand the sophisticated tests and scope techniques presented in full in the later chapters.

Clayton L. Hallmark

Contents

Functional Basics

An oscilloscope is a measuring instrument capable of measuring a wide variety of rapidly changing electrical phenomena, even phenomena occurring only once and lasting for a fraction of a millionth of a second.

The oscilloscope graphs changes in voltage with time. The amplitude, or strength, of the voltage is graphed along a vertical axis, and the length of time graphed along a horizontal axis. Because the graph of a voltage often takes the form of a wave, the graph is often called a **waveform.**

The ordinary moving-coil type of meter commonly used to measure ac and dc is somewhat limited in usefulness, since it only indicates a **value** of voltage or current. In the case of ac voltages and currents, for example, it gives the effective, or dc equivalent value, but no indication of the waveform. Since the waveform has a definite effect on the meter reading, the meter reading is accurate only if the waveform is the same as the meter is calibrated for. This means the meter reading is accurate only if the voltage or current being measured has the same waveform as power-line voltage and current. However, many of the voltages and currents in electronic equipment do not have this waveform, and an oscilloscope is required for their measurement.

By the same token, a simple meter becomes quite useless in the case of a pulsating dc consisting of an ac voltage superimposed, or riding, on a dc voltage. Such voltages are common in electronics, and their measurement also requires an oscilloscope.

Often it is impossible to tell whether or not a circuit is operating properly by merely checking the values of the voltages and currents in the circuit, even when they can be

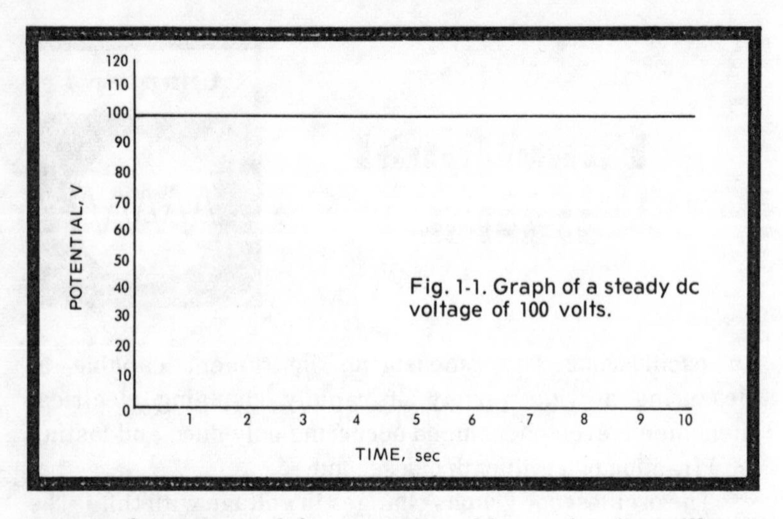

Fig. 1-1. Graph of a steady dc voltage of 100 volts.

accurately measured by a meter. Many circuits will not operate properly unless the ac voltages or the pulses involved have the correct **waveform**. For example, when the sound coming from a speaker is distorted, it is because the audio waveform has been changed in some circuit. The ac signal may have the proper amplitude (strength) and frequency, but it does not have the desired waveform. This intolerable defect would not be indicated by a voltmeter, but it would be indicated by an oscilloscope.

The use of the oscilloscope in waveform observation is only a small part of what it is capable of doing. The oscilloscope is the most useful and versatile of all electronic test instruments. It can measure voltage, current, time, gain, frequency, phase, hum, ripple, etc. But it is useful only if you understand it and it's versatile only if you really know how to use it. When you have finished this book, you will understand and really know how to use the oscilloscope. In order to more fully understand the usefulness of the oscilloscope for measuring ac voltages, we shall review the principles of ac and dc.

AC AND DC

Direct current (dc) is current that flows in one direction only. Its amplitude, or strength, may vary but not its polarity,

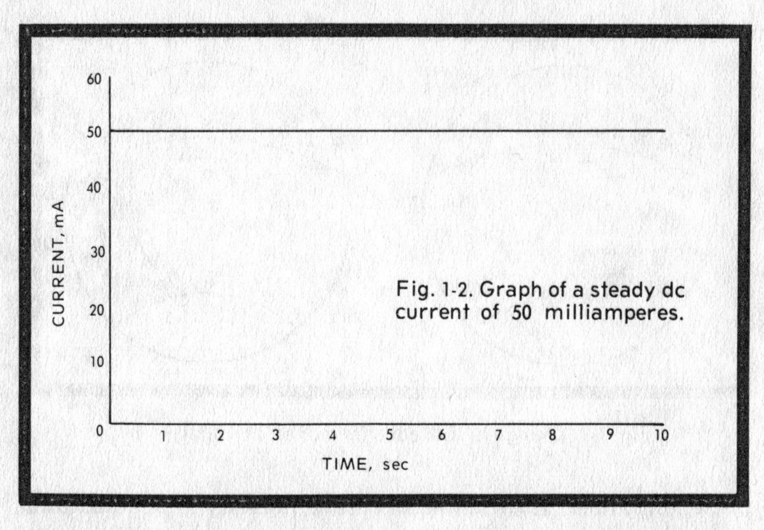

Fig. 1-2. Graph of a steady dc current of 50 milliamperes.

or direction. A source of dc, such as a battery or a dc power supply, can supply different amounts of current, depending on the power rating of the dc source and the resistance of the load connected across the source. If the resistance of the load should increase, for example, the current will decrease. Its direction, however, will always be the same, from the negative source terminal to the positive one. The direction will be the same for any length of time.

The graph of a steady dc potential of 100 volts appears in Fig. 1-1. Notice that time is marked off along the horizontal axis of the graph and that voltage is marked off along the vertical axis. A graph of the output current of a constant-current source delivering 50 mA is shown in Fig. 1-2. Again time is measured horizontally, and current is measured vertically. For both illustrations, the value of voltage or current is the same over the entire 10 second period. The value would in fact be the same for any length of time, unless some change were made in the source or the load. In both cases, it is sufficient to measure the value of voltage or current only once and be able to accurately state the value, not only at the time of measurement, but for all times.

Ac is another story altogether. It is constantly changing both in amplitude and polarity. A single measurement at a particular instant in time will not suffice to tell the amplitude for any other instant, only for the instant it is made.

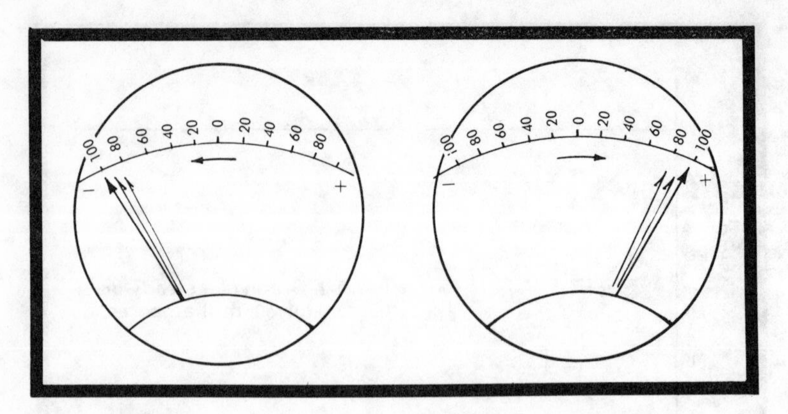

Fig. 1-3. Change is characteristic of an ac voltage.

The manner in which ac reverses polarity is regular and periodic. A 60 Hz voltage (or current), for example, increases from zero to a maximum in the positive direction, falls back to zero, continues down to a negative maximum and returns to zero again, 60 times a second. If we were to use a zero-center ac voltmeter to measure the 60 Hz voltage (Fig. 1-3), the needle would swing from center to the left, back through zero to the right, and back to zero again, 60 times per second. Of course, no meter exists whose needle can deflect so rapidly. Even if it did, the needle would move so fast that the human eye could not follow it. Thus, it is impossible to tell the direction much less the value of an ac voltage at any particular instant, using a meter.

Fig. 1-4 shows graphically the way an ac voltage varies with time. The horizontal line represents **elapsed time**. As time passes, from starting time A to time C, the voltage rises from 0 volts at A to 100 volts at B, and then falls back to zero. It then increases again, but this time in the opposite direction, until it reaches −100 volts at D. Finally it returns to 0 volts at point E. This completes one cycle. From point E to point I the cycle repeats itself. Using an oscilloscope, it is a simple matter to observe these variations we have been talking about, but using a meter it would be impossible.

The voltage at any instant, at A or B in Fig. 1-4, for example, is called an **instantaneous** voltage. Ac meters do not respond to instantaneous voltages. They indicate instead a

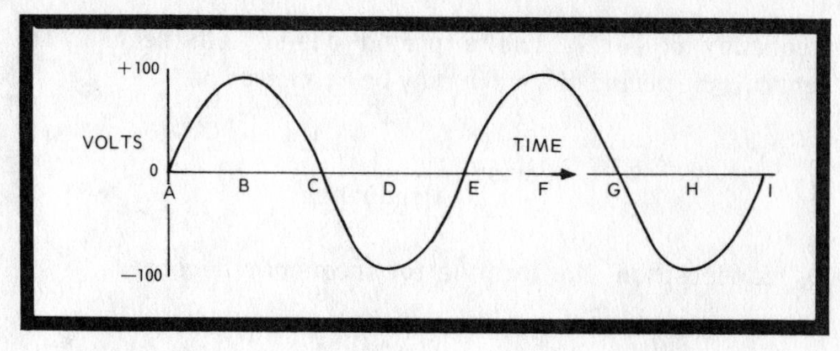

Fig. 1-4. One cycle of an ac voltage, as it actually appears in time.

sort of mean value—an rms, or effective value—which tells how large a battery would be required to supply the same amount of **power**. This is important, but it says little about the nature of the **variation** of an ac voltage. Only the oscilloscope can show the ac as it changes at each and every instant of its cycle.

Because ac cycles resemble waves when shown on a graph (or on an oscilloscope), they are often referred to as waves, and their graphical representations as waveforms.

FREQUENCY AND PERIOD

In the light bulbs, electrical appliances, and electrical machines in homes, offices, and factories around the country, the ac supplied by power companies is reversing itself in direction of flow 120 times per second. This means that the current is alternating directions 120 times per second, or, in other words, is operating at 120 alternations a second. Since a cycle consists of two alternations, the current is completing 60 cycles per second. This is its **frequency**, 60 cycles per second, or in modern terminology, 60 hertz (Hz). Frequency is defined as the number of complete cycles that occur during one second of time.

The period of time required for the completion of one cycle of a given ac current, the time between A and E in Fig. 1-4, is the **period** of the current or voltage, thus, a current that has a frequency of 10 Hz (goes through 10 cycles each second) has a period of one-tenth second. Similarly, a current that has a frequency of 60 Hz has a period one-sixtieth second. In general, the period of a wave may be expressed as

$$\text{period} = \frac{1}{\text{frequency}}$$

By transposition, the formula for frequency becomes

$$\text{frequency} = \frac{1}{\text{period}}$$

In both formulas, period is expressed in seconds, and frequency in hertz.

HARMONICS

The integral (whole number) multiples of any given frequency are known as **harmonics** of that frequency. The original frequency is called the **fundamental**. Thus, if we start out with 100 HHz, **that** is the fundamental; and its harmonics are 200 Hz, 300 Hz, 400 Hz, 500 Hz, etc. The harmonics are numbered according to which multiple of the fundamental they represent. Thus the second harmonic of 100 Hz is 200 Hz, and the third harmonic is 300 Hz, etc. By this reasoning, 100 Hz could be called the first harmonic, but usually it isn't; instead, we speak of the fundamental. Just because the first harmonic is rarely spoken of, don't forget that it exists. And don't make the mistake of confusing the first harmonic with the second harmonic, which is actually twice the first harmonic, or fundamental. That is a common mistake.

The harmonics of a frequency are referred to as **even** or **odd harmonics**, according to whether they are even or odd multiples of the fundamental. Thus, 300 Hz, 500 Hz, 700 Hz, etc., are odd harmonics of 100 Hz. Similarly, 200 Hz, 400 Hz, 600 Hz, etc. are even harmonics of 100 Hz.

The significance of harmonics in the study of waveforms is that various waveforms are made up of different harmonics of the basic waveform, which we will discuss next.

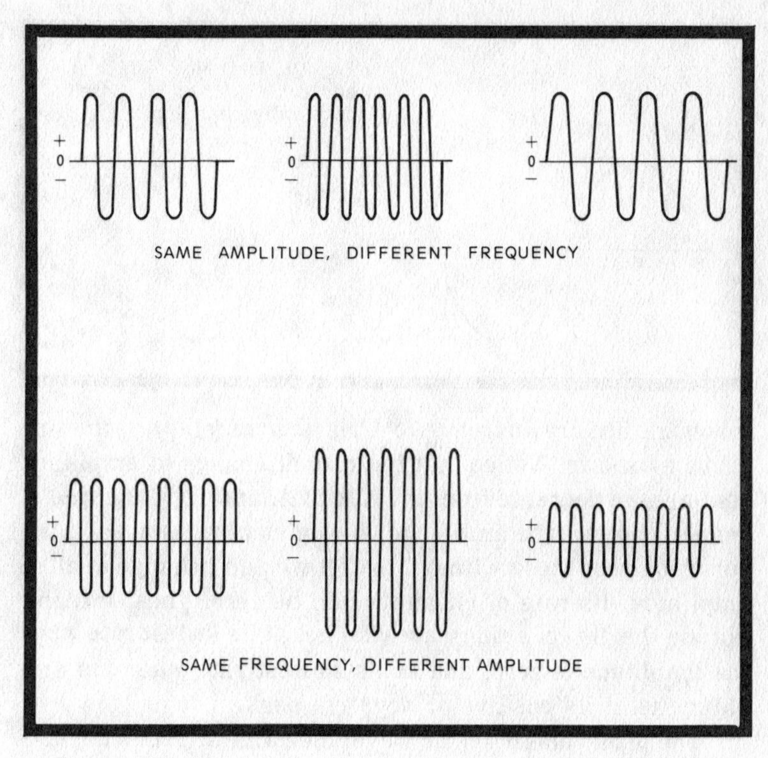

Fig. 1-5. All of the above waves are different, but all are sine waves.

THE SINE WAVE

The most common ac waveform is the sine wave, shown in Figs. 1-4 and 1-5. This waveform is typical of power-line ac and of the output of most af and rf signal generators.

All of the waves in Fig. 1-5 are different, but all are sine waves nonetheless. They all have certain characteristics that distinguish sine waves. All of the curves rise smoothly from zero amplitude to peak amplitude and decline smoothly back to zero amplitude, alternately in the positive and negative directions. Also, the amplitude of each curve changes fastest when its amplitude is zero, and changes slowest when its amplitude is maximum. Fig. 1-6 illustrates this. Notice that at point A in Fig. 1-6 the amplitude of the sine curve is nearly zero. It is, however, increasing with time. If it continued to increase at the same rate as it does at A, the curve would

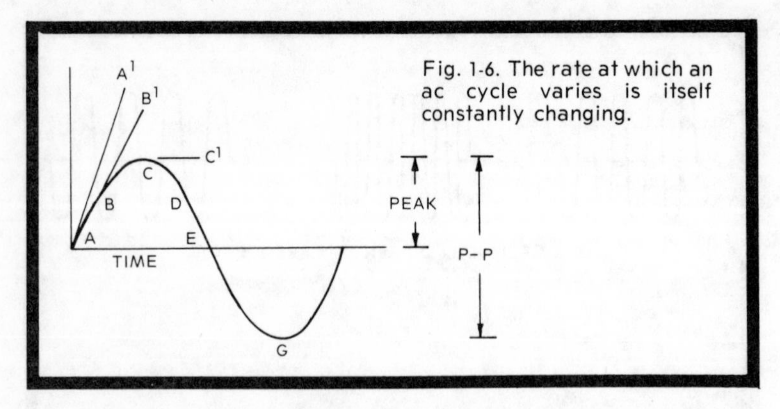

Fig. 1-6. The rate at which an ac cycle varies is itself constantly changing.

follow the line drawn from A to A', instead of following the sine curve as shown. Notice that the rate of change in amplitude continues to decrease from the rate at A, until at C the rate of change (but not the amplitude) becomes zero. That is, if the curve followed the line from C to C' it wouldn't change at all in amplitude. Its rate of change would be zero. Thus, the amplitude (height) of a sine wave changes at its fastest rate when the amplitude is zero, and at its slowest rate when the amplitude is at its positive or negative peak.

The **peak value** of the ac wave, the value at points C and G in Fig. 1-6, is equal to 1.414 times the rms value, which would be measured by an ac voltmeter. The **peak-to-peak value**, the vertical distance between points C and G in Fig. 1-6, is twice the peak value, or 2.828 times the rms value. Both the peak value and the peak-to-peak (p-p) value of a sine wave can be determined with an ordinary ac voltmeter. The peak and p-p values of other waveforms must generally be determined with an oscilloscope.

Another important point about sine waves is that they are considered basic. They are the standard by which other waves are judged. Other waves are considered as combinations of sine waves. The sawtooth, square wave, and trapezoidal waves we will look at soon are each considered to be a different combination of many sine waves of different amplitude and frequency.

The selection of the sine wave as the basic wave was not arbitrary. There is much precedent for it in nature. In electronics too, there is good reason to consider the sine wave

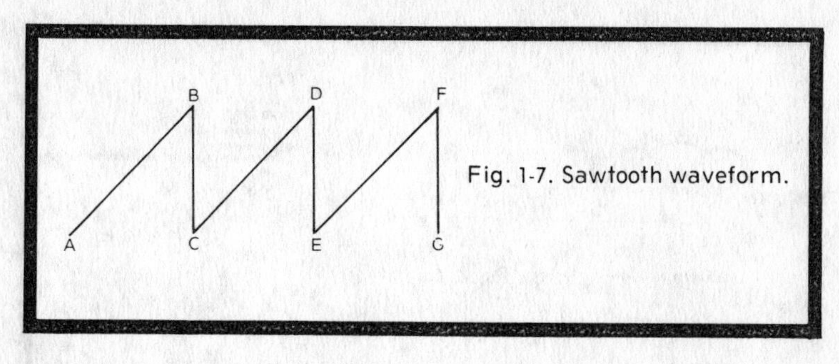

Fig. 1-7. Sawtooth waveform.

basic. It is the only waveform that will retain its shape after passing through a network containing capacitors or inductors. Other waveforms are distorted by such networks.

THE SAWTOOTH WAVE

The sawtooth wave is possibly the second most important waveform. This wave, named for its resemblance to the teeth on a saw, is widely used in oscilloscopes and other measuring instruments, as well as in television. The ideal sawtooth has a linear (straight-line) rise, as shown by the lines AB, CD, and EF in Fig. 1-7. It also has a quick, abrupt decline, as shown by the lines BC, DE, and FG. Like most ideals, the ideal sawtooth is not quite realized in practice. The rise is not perfectly linear nor the drop perfectly abrupt.

The sawtooth is used in oscilloscopes and televisions to sweep a luminous dot across the crt screen, creating a display or picture.

THE SQUARE WAVE

The so-called square wave is usually rectangular as in Fig. 1-8, rather than square. This wave is distinguished by its

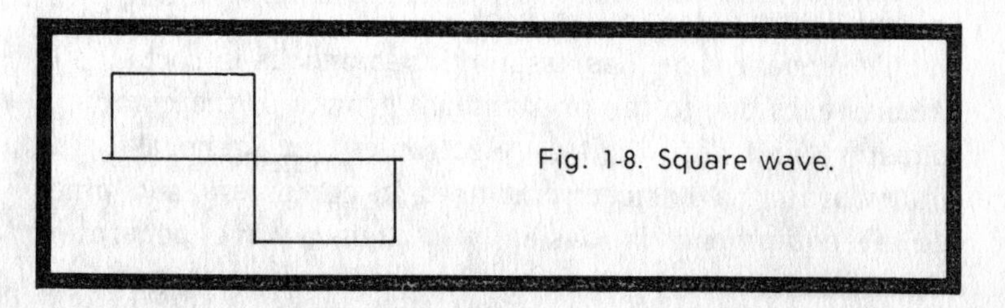
Fig. 1-8. Square wave.

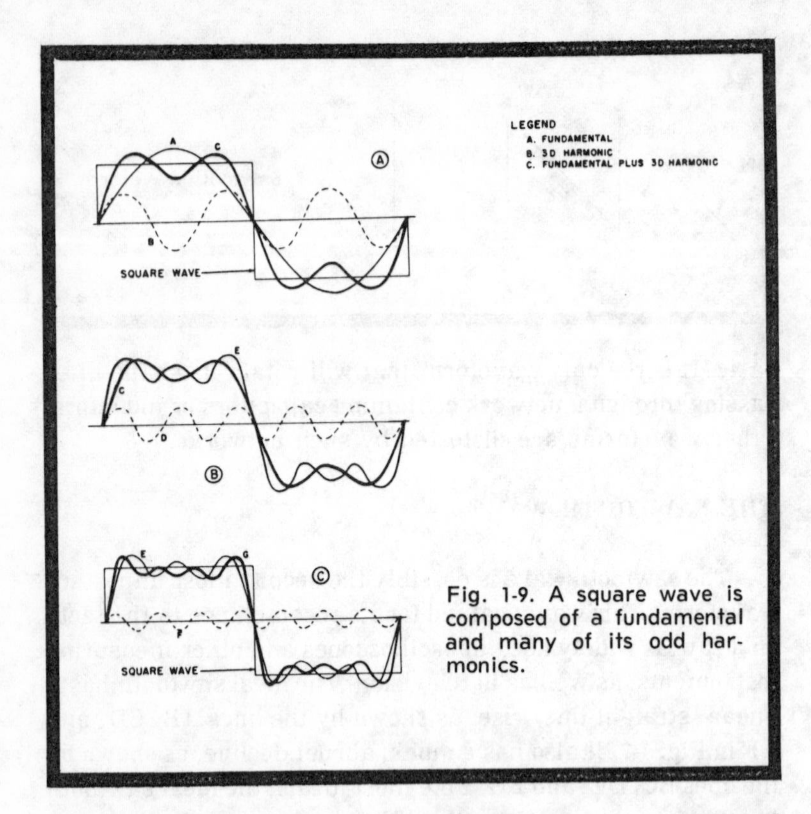

Fig. 1-9. A square wave is composed of a fundamental and many of its odd harmonics.

steep sides, which indicate an almost instantaneous rise and fall. It is also distinguished by a flat top.

Even though it looks simple, this wave is actually made up of a large number of sine-wave harmonics. Fig. 1-9 shows how the square wave is built up from a fundamental sine wave and its odd harmonics. Note that as more odd harmonics are added, the composite wave gets nearer and nearer to its ideal shape. A good square wave has many harmonics and is thus useful for testing the response of equipment to a wide range of frequencies. We will show you how square-wave testing is accomplished, later in this book.

The square wave has assumed tremendous importance in recent years due to the phenomenal growth of the computer industry and of digital electronics in general. The multivibrator, a basic circuit used in computers and other digital equipment, is basically a square-wave generator. Multivibrators are, by the way, also used in oscilloscopes.

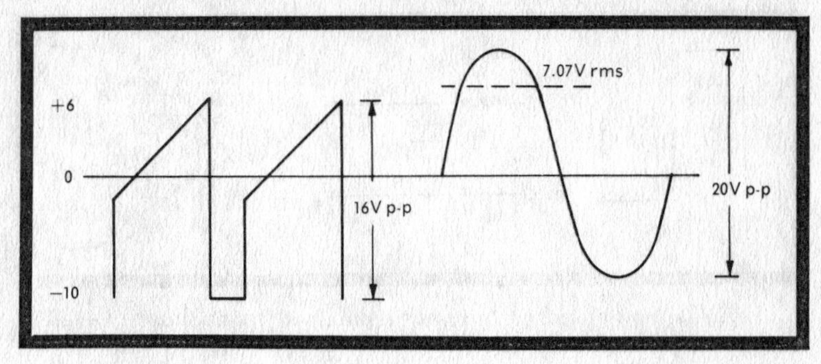

Fig. 1-10. A trapezoidal wave, and a sine wave shown for reference.

OTHER WAVES

The trapezoidal waveform in Fig. 1-10 is a combination of a square wave and a sawtooth wave. It is actually a sawtooth wave riding atop a square wave.

Notice that the positive peak voltage of the trapezoidal wave is different from the negative peak. The positive peak is 6 volts, while the negative one is 10 volts. In order to measure such an asymmetrical voltage with a scope, it is compared to a sine wave, which is symmetrical. As illustrated in Fig. 1-10, the sine wave gives a zero reference for the asymmetrical trapezoidal wave.

Trapezoidal waveforms are encountered in television sweep circuits at the output of the vertical and horizontal oscillators.

When a square wave is passed through the circuit in Fig. 1-11 the **differentiated** waveform shown in Fig. 1-11 results. The square wave is converted into a series of voltage spikes. When a square wave is applied to the circuit in Fig. 1-12, another series of pulses results. These constitute an **integrated**

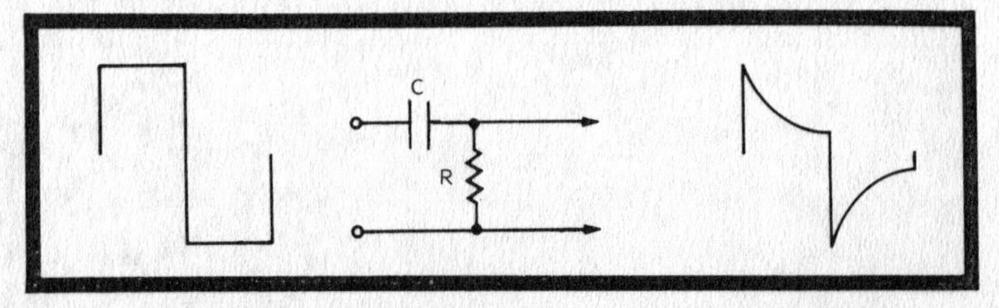

Fig. 1-11. How a differentiated wave is obtained from a square wave.

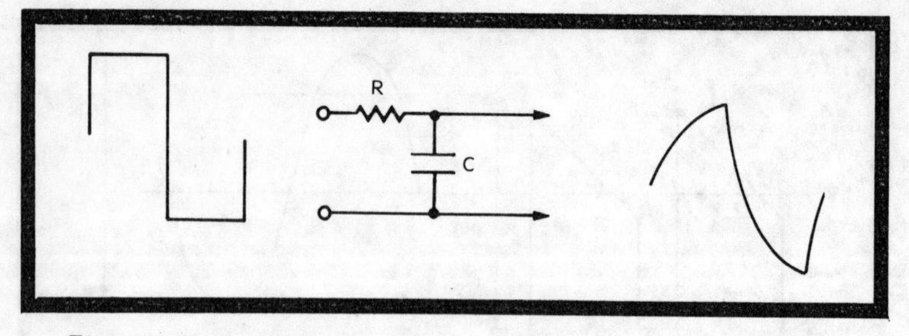

Fig. 1-12. How an integrated wave is obtained from a square wave.

waveform. Both differentiated and integrated waveforms are encountered in TV sync circuits. Notice that the differentiated waveform is a type of trapezoidal waveform and the integrated waveform is a type of sawtooth.

The Cathode-Ray Tube

The cathode-ray tube (crt) is a special type of vacuum tube in which the electrons emitted by the cathode are focused into a narrow beam, accelerated to a high velocity, and then caused to strike a chemically coated screen. The screen fluoresces, or emits light, wherever it is struck by the beam.

The usefulness of the crt lies in its ability to produce a visible trace of light on a screen. By causing the trace to follow certain patterns, we have a means of **visually** representing voltages in all types of electronic circuits. Being almost weightless, the electron beam can move about the screen quite rapidly in response to voltage variations, and thus can display variations lasting but a fraction of a microsecond. The crt is the most important part, the basis, of an oscilloscope. Due to the rapid movement of the electron beam, the oscilloscope may be used to observe voltages at very high frequencies.

Besides being used in oscilloscopes, the crt is used as a display device in many other types of electronic instruments. In television, it is used in transmitting as well as receiving equipment. The camera tube of a TV camera is a type of crt that changes light variations into electrical variations. Cathode-ray tubes are also used in radar, where the information picked up by the radar system appears on the face of a special crt called a radarscope. In a **loran** navigational system on a ship, the position of two blips on a special crt called a **loran screen** enables the operator of the system to determine the location and bearing of his ship.

Cathode ray tubes are used in many applications, from automobiles to zoological research. Many of the most important applications will be discussed in this book.

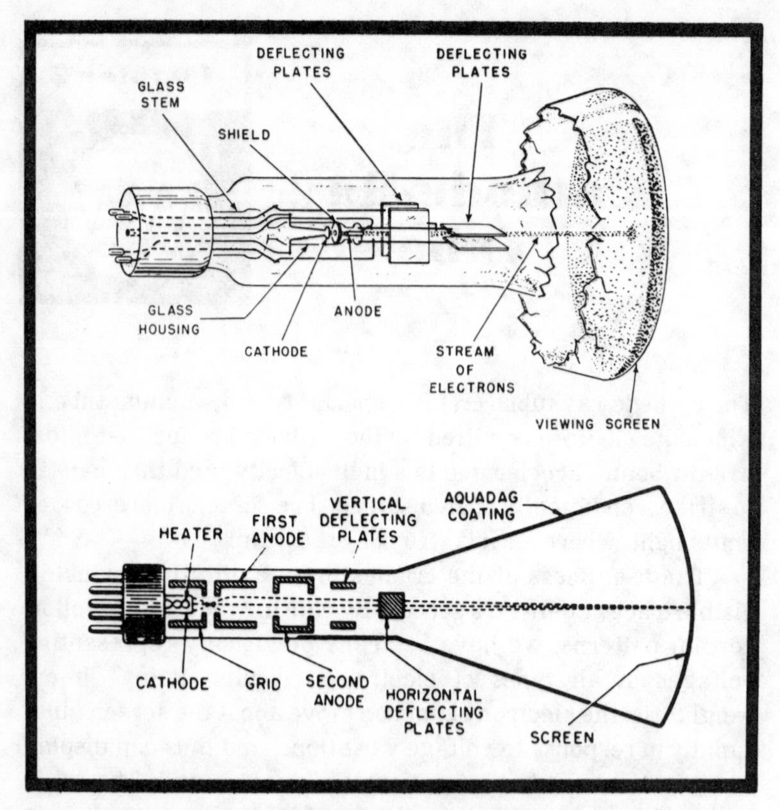

Fig. 2-1. The electron beam emitted by the cathode flows through the grid "cage" toward the phosphor-coated face of the tube. The first anode is the focusing element, and the second anode is the beam-accelerating element. The vertical and horizontal deflection plates allow the beam to be directed toward any part of the screen.

CONSTRUCTION OF THE CRT

The ordinary vacuum tube is concentric; that is, the cathode is a cylinder, and the plate is a cylinder that surrounds the cathode. Electrons are emitted radially from the cathode to the plate, spreading out as spokes spread out from the hub of a wheel to the rim. In the crt, however, the electrons are acted upon by certain internal elements that focus the electrons into a beam, a **cathode ray**. The path of the electron beam, or cathode ray, is shown in Fig. 2-1.

Electrons in the beam are accelerated to a high velocity and directed so as to strike the inside of the crt face. The face

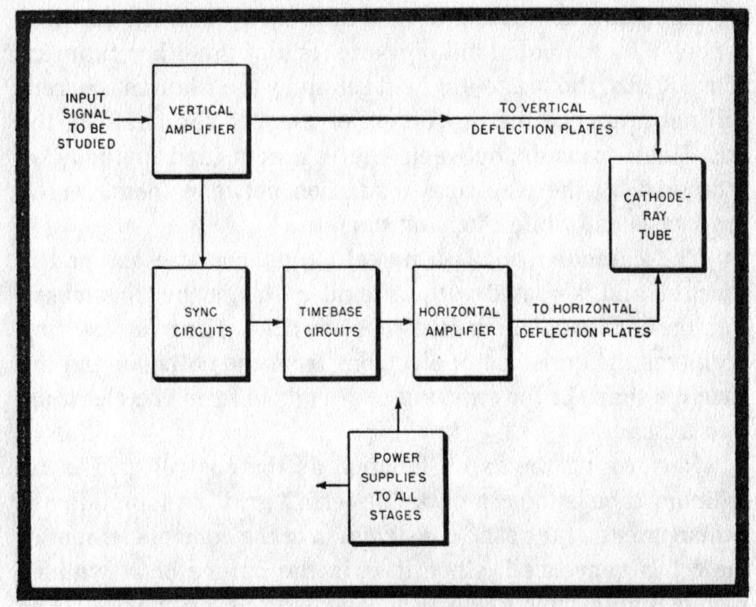

Fig. 2-2. Block diagram of a conventional oscilloscope.

of the tube is coated with a "phosphor" chemical, which emits visible light when bombarded by the electrons. A small point of light will thus be produced wherever the electron beam strikes the face of the crt. This is called **fluorescence**. As the beam is moved around the crt face, or screen, a line is traced out, because the phosphor has the property of continuing to emit light for a short time after being bombarded. This property is called **phosphorescence**. Shortly, we will discuss the means by which the beam is moved about the screen of the crt. Fig. 2-2 shows the function of the crt.

ELECTRON GUN

The portion of the crt that produces and directs the electron beam is the **electron gun**, an electrode assembly mounted in the neck of the tube. Fig. 2-1 shows that the electron gun is made up of a filament (heater), a cathode, a grid, and a first and second anode.

The cathode is indirectly heated by the **filament**, which is wound in the form of a double spiral. In the double spiral arrangement, the magnetic field built up around one spiral is canceled by the equal but opposite field of the other filament spiral. Thus, the magnetic field set up by the filament current will not influence the movement of the electron beam in the crt. Heat transfer between the filament and cathode is promoted by the electrical insulation between them, which conducts heat while blocking current.

The **cathode** itself is a nickel cylinder whose one end is concave and is coated with an oxide. This is the "business" end, the end from which electrons are fired. The oxide coating promotes the emission of electrons from the cathode, and the concave shape of the emitting end tends to form the electrons into a beam.

Performing the same function as the control grid in an ordinary tube is the crt **grid**. The term "grid" is something of a misnomer as far as the construction of the control element of the crt is concerned, since it is in the nature of a cylinder rather than a wire mesh. But it acts as an ordinary grid in controlling the rate at which electrons leave the cathode. It also concentrates the emitted electrons into a tight beam. Fig. 2-1 shows that the electrons emitted from the cathode are confined within the walls of the grid, and that the only way they can leave is through the small hole in the center of the grid. Since the grid is maintained at a negative voltage and since it surrounds the emitted electrons, it tends to force the electrons toward its own axis as they pass through. Controlling the intensity of the electron beam is, however, the main function of the grid. The influence of the grid extends for only a short distance, where the beam comes to a point. Here the first anode takes control.

The first anode is also called the **focusing anode**. Because it is positive, the first anode attracts the electrons leaving the grid. Because it is cylindrical, the anode exerts the same pull on all sides of the electron beam passing through it. The net effect is to focus the electron beam into a pencil of light with very nearly parallel sides. Usually, the first anode has a positive potential on the order of that used on the plates of ordinary tubes. This voltage is variable by a focusing control, and it determines the **thickness** of the electron beam. Take care not to confuse **thickness** with **number of electrons**, which is controlled by the **grid**.

Next in the path of the electron beam is the **second anode**, or accelerating anode. As its name implies, the accelerating anode has the job of speeding the electrons up as they move toward the screen. To obtain a very high velocity, the second anode is maintained several thousand volts positive compared to the cathode potential. A speed of about ten thousand miles per second is achieved by the electrons as they leave the second anode.

Electrically connected to the second anode is an **Aquadag** coating on the inside surface of the crt (Fig. 2-3). This coating is made of graphite, a conductor, and provides the means by which the electrons in the electron beam return to the high-voltage power supply and the cathode, thus completing the circuit. When the beam bombards the screen, secondary electrons knocked off are picked up by the highly positive Aquadag coating. If electrons from the beam collected on the screen and none were bumped off, the screen would soon start to repel the electrons of the electron beam and prevent proper operation of the crt. In addition to capturing the rebounding electrons, the Aquadag coating shields the electron beam from any stray magnetic fields that might be present.

Deflection System

Since the electrons that comprise the beam are negative, the beam is repelled by a negative charge or attracted by a positive charge. Thus the beam can be deflected, or bent, by passing it between charged plates. By proper positioning of the plates, and by varying the voltage on them, we can move the beam up and down and side to side on the screen, drawing a picture much as we would with a pencil.

The deflection system used in an oscilloscope is an **electrostatic** deflection system. Taken together, the plates between which the beam is passed act much like a capacitor (See Fig. 2-3.) When a voltage is applied to them, that is, when charges are placed on them, an electric field exists in the space between the plates. The beam will be attracted to the positive plate and repelled by the negative plate.

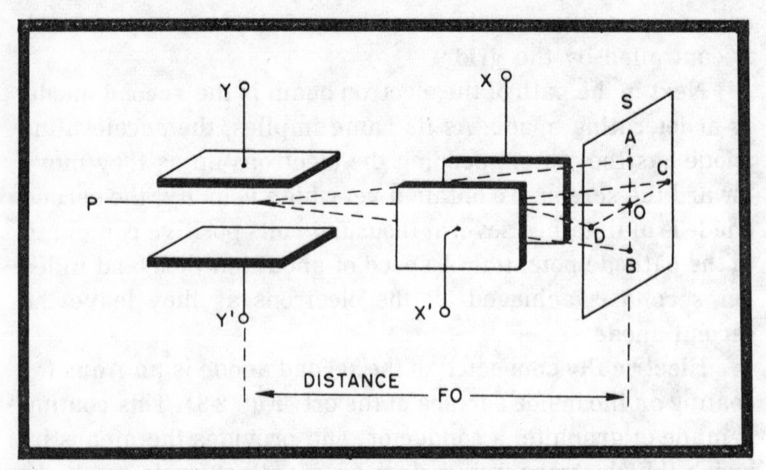

Fig. 2-3. Vertical and horizontal deflection plates. Note that horizontally oriented plates cause vertical deflection and vertically oriented plates cause horizontal deflection of beam. Vertical deflection plates (YY') are nearest cathode, so angle of deflection between distance FO is lower than angle for horizontal plates (XX')

In Fig. 2-4, two plates are mounted horizontally and a dc voltage is applied, making the upper plate positive and the lower plate negative. Thus the beam is bent upward so that it will strike the face of the crt at a point above the center of the screen. Since the horizontally mounted plates cause the electron beam to deflect vertically (up and down), they are called the **vertical deflection plates.** We emphasize that the designation of these plates as "vertical" relates to their effect rather than their orientation in the crt. Fig. 2-5(a) shows where the beam would strike the screen if there were no voltage on the plates. How the situation in Fig. 2-4 will appear from the front of the screen is shown in Fig. 2-5(b).

If we were to double the potential between the plates in Fig. 2-4, the beam would be deflected twice as much and the dot on the crt screen would move twice as far, as in Fig. 2-5(c). Hence the amount of deflection varies according to the relative voltage amplitudes and polarities on the plates. For example, if we were to apply the same voltage as in Fig. 2-5(c), but in the opposite polarity, the electron beam would be deflected downward, and the luminous dot on the screen would appear as in Fig. 2-5(d). Thus, the direction of the beam's deflection depends on the polarity of the voltage on the plates.

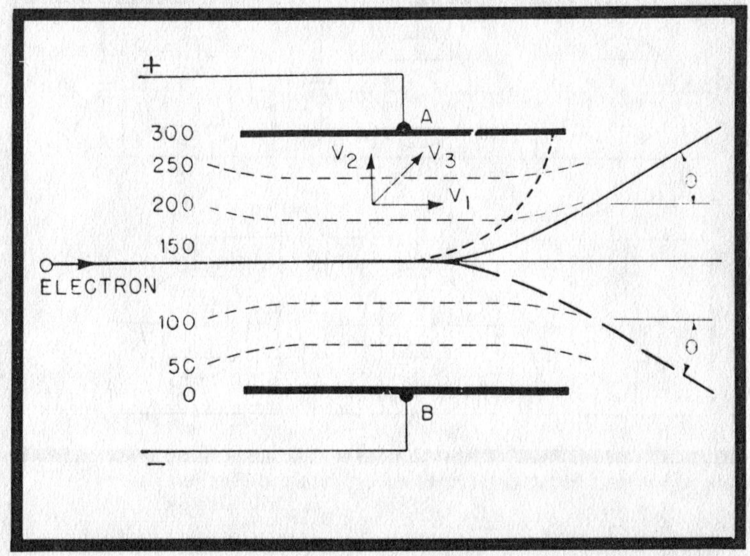

Fig. 2-4. Deflection of electrons between parallel plates. If excessive potentials are used, electrons will strike edges of plates (broken line).

The orientation of the deflection plates is shown in Fig. 2-3. Since the horizontally mounted plates cause the electron beam to deflect vertically (up and down), they are called **vertical** deflection plates. The designation of "vertical" refers to their effect rather than their orientation in the crt.

Fig. 2-5A shows where the beam would strike the screen if there were no voltages on any of the deflection plates. Increasing the positive potential of the Y plate (Fig. 2-3) causes increased attraction by that plate, with the result that the electron beam is drawn toward it. The effect is as shown in Fig. 2-4.

By alternating the polarity on the plates and moving the beam up and down very rapidly, a line of light can be made to appear on the screen. This is due to two different effects; the persistence of the screen and the persistence of vision. The persistence of the screen, which we will delve into in greater depth later, is due to the previously explained phenomenon of phosphorescence. Persistence of vision is the human tendency to see a thing for a short time after it has disappeared from view. Thus, when the beam moves rapidly up and down on the crt screen, we see a continuous line instead of a series of dots.

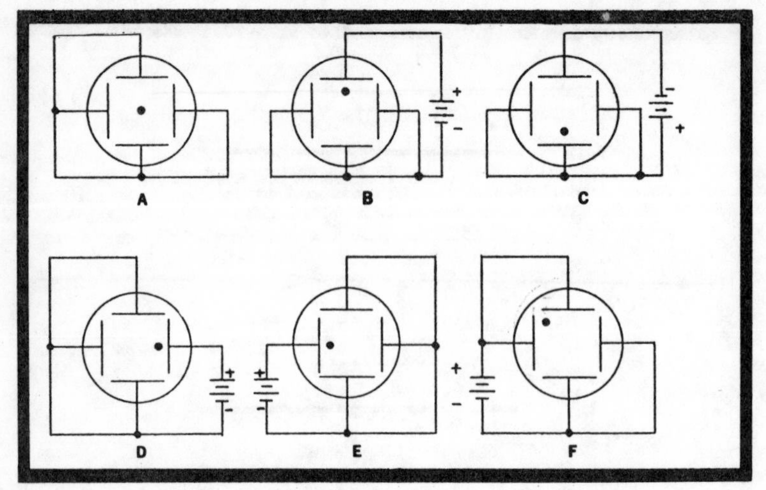

Fig. 2-5. Simplified crt beam deflection.

Focusing Action

We know how to make an electron beam move up and down and from side to side, but how do we focus the beam so that it results in a sharp and tiny dot on the screen? As shown in Fig. 2-6, the first, or focusing, anode has a high potential—in this case, 1200 volts. The accelerating anode's potential here is 2000 volts, or 2 kV. Because of the voltage difference between the two anodes, there is a strong electrostatic field set up. This is represented in Fig. 2-6 as a series of curved lines. The strength of the field determines the point at which the beam lines will converge. The electrons entering this electrostatic "lens" have two forces acting upon them: the accelerating force because of the second anode's attraction, and the converging force because of the field's intensity between the anodes.

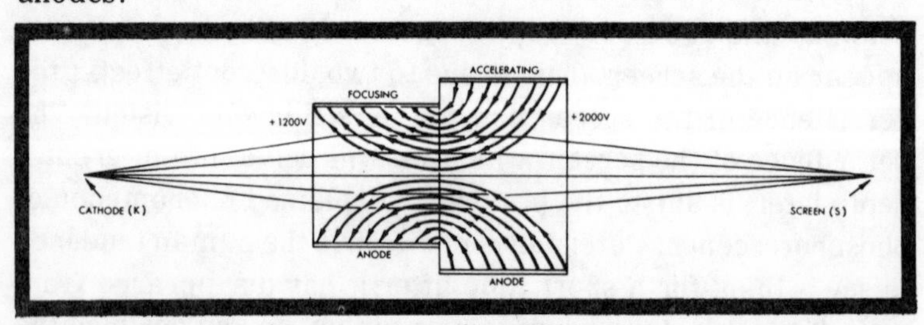

Fig. 2-6. Focusing action in crt. The focusing anode is the first anode, and accelerating anode is the second.

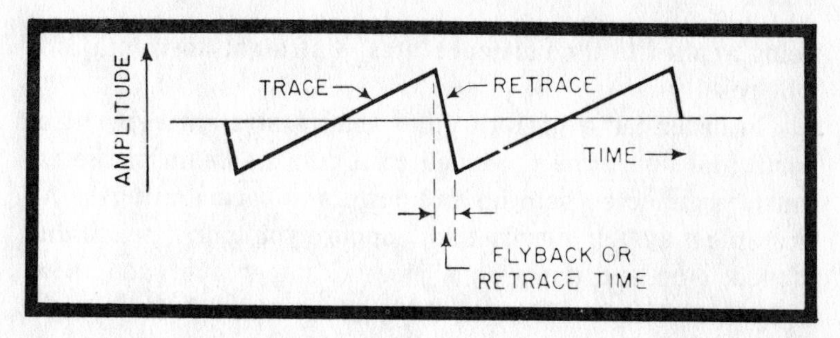

Fig. 2-7. The electron beam traces out a picture on the screen and the beam is swept across the screen while an ac voltage is applied to the vertical deflection plates. The vertical aspect of the display represents voltage, while the horizontal aspect represents time.

Display

If no voltage is applied to the vertical deflection plates, there will be no vertical deflection. If, however, a voltage is applied to the **horizontal** deflection plates at this time, the beam will be deflected horizontally. If the beam is deflected back and forth rapidly enough, a horizontal line will appear on the crt screen. The length of the line will be proportional to the strength of the ac voltage sweeping the beam back and forth.

Now let us assume that an ac voltage is applied to the vertical deflection plates and that it is causing a dot of light to move up and down on the crt screen. Assume also that at the same time a gradually changing dc voltage is applied to the horizontal deflection plates, and that its polarity is such as to cause the beam to move horizontally across the screen, from the extreme left to the extreme right. As the beam is moving from left to right, it will also be moving up and down due to the ac voltage applied to the vertical deflection plates. In this manner, the luminous dot on the screen will trace out a picture of the voltage applied to the vertical deflection plates, as in Fig. 2-7.

A sine wave will be traced out on the screen due to an ac voltage on the vertical plates only if there is a voltage applied to the horizontal plates to sweep the dot back and forth on the screen. If there is no voltage applied to the horizontal plates, a straight vertical line will result. On the other hand, if the beam is swept back and forth across the screen without any voltage being applied to the vertical plates, a straight horizontal line will result.

In order for a pattern other than a straight vertical or horizontal line to be produced on a crt, the beam of the crt must be deflected both up and down and back and forth. An example may help clarify this. Suppose you took a pencil and moved it up and down on a piece of paper over and over, without moving it either to the right or the left. A straight vertical line would result. On the other hand, suppose you moved your hand steadily to the right while moving the pencil up and down with your fingers. A wavy line would result. This is the way you would draw a sine wave, and it is the way a crt draws a sine wave. It moves a dot across the screen in two directions at the same time. In order to do so, it requires two varying voltages at the same time. The dc voltage that sweeps the beam across the screen in the horizontal direction is supplied by the circuitry of the oscilloscope. The ac voltage applied to the vertical deflection plates to move the beam up and down as it is moved across the screen is supplied from some outside signal source. This is the voltage that is measured or graphed on the crt.

The manner in which the electron beam is moved back and forth across the screen of the crt is rather similar to the way in which the carriage of my typewriter moves back and forth as I write this. I start writing at the left side of the page and proceed gradually (very gradually, in fact) across the page until I reach the right side. Then I hit the carriage return and return to the left side of the page to begin writing the next line. The speed with which the carriage returns to the left side of the page is much greater than the speed with which I type across the page, of course. A similar action is applied to the electron beam by the voltage supplied by the oscilloscope to its horizontal deflection plates. The electron beam is swept gradually and steadily to the right until it reaches the extreme right side of the screen, and then it is swept back, almost instantly, to the left side. We see the pattern traced out by the electron beam on the screen as it moves from left to right, but we normally do not see the beam return to the left, because it returns so rapidly. The time required for the beam to return to

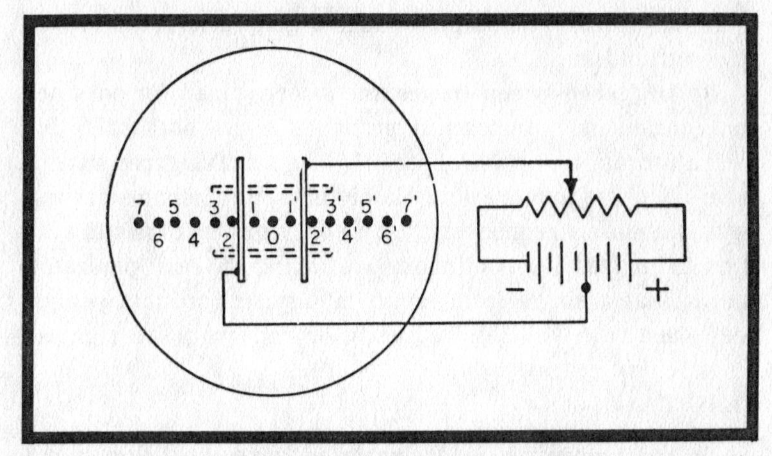

Fig. 2-8. At the start of the sweep, the beam is at the left side of screen. As sawtooth voltage increases, time goes by; and as time goes by, beam is deflected further and further to the right. The sweep ends when the sawtooth abruptly stops and returns to zero voltage.

the left side is called the **retrace**, or **flyback time**. Also, in many oscilloscopes, a highly negative **blanking pulse** from a multivibrator is applied to the grid of the crt during retrace to shut off the beam. This insures that nothing is seen on the screen during retrace, which is as it should be.

In order to cause the electron beam to sweep gradually across the screen to the right and then return abruptly to the left side of the screen, a voltage that increases gradually from zero volts to a maximum voltage and then returns abruptly to zero volts is required. The waveform that has this characteristic is the sawtooth waveform shown in Fig. 2-7.

During the period of time that the voltage is gradually increasing, the electron beam of the crt is being slowly swept across the screen from left to right, as shown in Fig. 2-8. When the voltage suddenly falls to zero, the electron beam is snapped back to the left. This process may be repeated over and over, and the beam repeatedly swept across the screen. The voltage applied to the horizontal deflection plates is often called the **sweep voltage**, because it sweeps the beam across the screen.

In some oscilloscopes the sweep generator runs continuously, and its frequency is adjusted so that it corresponds to the frequency of the signal to be observed, or to a submultiple thereof. The output of such a generator is a series of sawtooth pulses.

In **triggered-sweep scopes**, the sweep generator does not run continuously. Instead, it produces a sawtooth pulse for each alternation or pulse of the input signal. Triggered sweep, to be discussed in more detail later, is used to permit viewing signals whose frequencies vary, or signals consisting of random pulses. A continuously running sweep generator operates at a single frequency it is adjusted to, and cannot keep pace with varying-frequency or random-pulse signals.

TIME BASE

In most cases, it is desirable to have the beam move across the screen at a uniform rate of speed. In considering any voltage wave, you should bear in mind that successive points on the wave are actually different values of voltage following each other in time. Also in most cases, succeeding voltage waves are all identical to one another. When such a wave is shown on an oscilloscope, the succeeding points are displaced in space, from left to right. Thus, the horizontal sweep of a scope is really a time scale, or **time base**.

The term **time base** is used in connection with scopes having either continuously running sweep generators **or** with triggered sweep. The horizontal calibration of scopes with continuously running sweep generators is actually in terms of **frequency**, since it is the frequency of such generators that is adjusted. Such scopes, however, are still considered to have a time base. The newer, triggered-sweep scopes have sweeps calibrated in terms of a **direct** unit of time for a given distance of spot travel across the screen, hence the term **time base** is more applicable to these scopes than the earlier types with continuously running sweep generators. The time base of a triggered-sweep oscilloscope is specified in terms of **time per division** (time / div). That is, the time base is specified in terms of how long it takes the spot of light on the screen to move a distance equal to a side of one of the squares marked off on the screen, or on an appliqued **graticule**.

Whether the sweep is triggered or not, it must be linear, or uniform in speed. In order for the sweep to be linear, the sweep voltage amplitude must change at a linear rate as it sweeps the beam across the screen. This means that the ramp of the sawtooth (Fig. 2-7) must be straight, or linear, instead of curved.

If the beam is not deflected linearly across the screen, the time base will not be uniform throughout the entire pattern. That means the time base will be distorted. Instead of the two alternations of the sine wave having the same width, one is compressed. This could occur if the beam moved across the screen from left to right at a slower rate during the one alternation than during the other.

FLUORESCENT SCREEN

The purpose of the fluorescent screen is to convert the energy of the electrons bombarding it into light energy. The inside face of a crt is coated with a phosphorescent material that continues to emit light for a short time after it has been bombarded by the electrons of the electron beam. This afterglow is called **persistence**, and it varies according to the screen material. Accordingly, the persistence may be designated as short, medium, or long persistence. The length of the persistence used in a particular scope depends on the application of the scope. General-purpose scopes have medium to long persistence. A medium-persistence phosphor is used in TV. In a radarscope, a long-persistence phosphor is used, since it may be necessary to see the trace or its slow motion for quite some time after the first instant of observation.

The color of the light on the screen as well as the persistence is determined by the type of chemical used to coat the screen. Willemite will produce green light, zinc oxide will produce blue, zinc beryllium will give yellow, and a combination of zinc sulphide and zinc beryllium will give white light.

The persistence desired in an oscilloscope screen depends on the frequency of the signals to be observed. For observing low-frequency waveforms, a longer persistence is desirable, since the crt must retain one trace for a fairly long time until

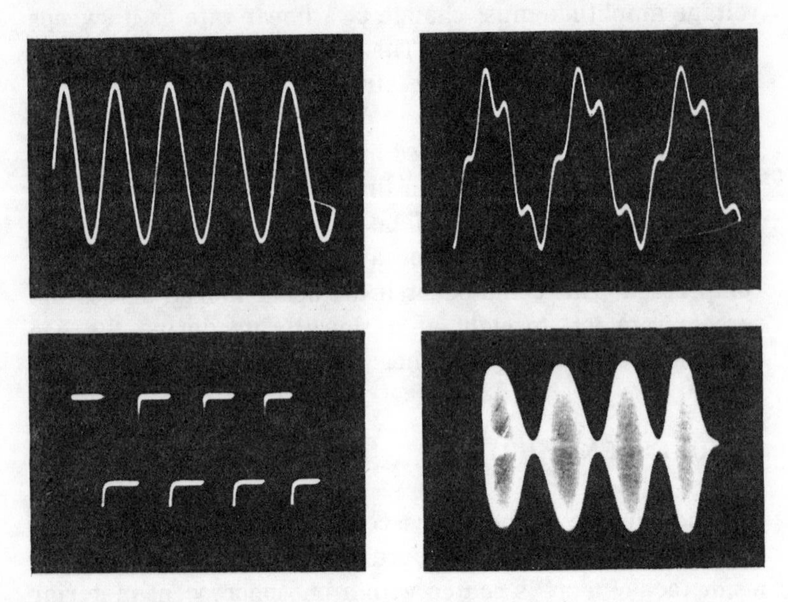

Fig. 2-9. Typical waveforms of low-persistent scope screens.

the next one comes along. On the other hand, for a rapidly changing image (high frequency), a short persistence is desired so that one image can fade before the next one comes along. Typical fast-fade displays are shown in the photos of Fig. 2-9.

The persistence of the screen can be determined from the type number assigned to the tube. The letter "P" with a number following it designates the type of phosphor used, and thus the color and persistence of the trace produced on the screen. For example, a 5UP7 type has a P7 phosphor. As Table 2-1 shows, the P7 phosphor has medium persistence. When the electron beam first strikes a spot on the 5UP7's P7 phosphor, a blue light is emitted (fluorescence). This changes to yellowish-green (phosphorescence). As the table shows, other phosphors have different fluorescence and phosphorescence characteristics, as well as different persistences. The characteristics of a given crt screen can be determined from the tube type number and the table. In scope catalogs, the catalog description indicates the phosphor normally supplied with each scope. However, for specific applications, it is often possible to specify another phosphor.

Table 2-1. Comparisons of screen phosphors and parameters.

Type of Phosphor	Fluorescence	Percent, Relative Luminance	Milliseconds for brightness to fall to 0.1 percent	Relative burn resistence	General
P1	Yellowish green	50	95	Medium	Replaced by P31 for most applications.
P2	Bluish green	55	120	Medium high	Good for hi and lo frequencies.
P4	White *	50	20	Medium high	Used in TV.
P7	Blue	35	1500	Medium	Slow fade
P11	Purplish blue *	15	20	Medium	For photo applications.
P15	Bluish green	15	0.05	Very high	For use in flying-spot scanner.
P31	Yellowish-green	100	32	High	General purpose

* Yellow-green phosphorescence

Human Eye Response

One important factor in selecting a phosphor is the color of the light output. The human eye is peaked in its response to light in the yellow-green region and falls off on either side, in the orange-yellow range or in the blue-violet range.

If the quantity of light reaching the eye is doubled, the brightness seen, or perceived by the eye is **not** doubled. According to Weber's Law, the brightness perceived is approximately proportional to the logarithm of the stimulus. The brightness must be increased 100 times before it seems to be doubled.

The term **luminance** used in Table 2-1 is the photometric equivalent of brightness. It is based on measurements made with an instrument having about the same response as the average human eye. The luminance information in the table is thus useful only when the phosphor is being viewed visually, and is not useful for photographic purposes.

Phosphor Burning

When a phosphor is bombarded by an electron beam having too great a current density, a permanent loss of phosphor efficiency is likely to result. The light output of the phosphor is diminished, and in extreme cases, the phosphor may be destroyed. Burning or darkening occurs when the phosphor is unable to dissipate the heat caused by electron bombardment fast enough to prevent the heat from building up to a destructive level.

Burning depends not only on the intensity of the electron beam, but also on the length of time it strikes a certain spot. The intensity of the electron beam is controlled by the intensity, focus, and astigmatism controls (to be discussed later). The length of time the beam excites a given spot depends on the setting of the time/div control (also to be discussed later).

Sweep Circuits (Time Bases)

The cathode-ray tube, or crt, is the heart of a scope, but alone it is useless. To operate, it needs at least a power supply to provide its electrode voltages, and a sweep generator to move the electron beam across the screen. Usually, one or two amplifiers are also required. Together, all these devices constitute an oscilloscope.

Power supply and amplifier circuits are used in almost all electronic equipment, but sweep generators are more specialized circuits found mainly in scopes and TVs. Because of this fact, and because there are so few books on scopes that cover the subject of **transistor** sweep circuits fully, we are devoting this chapter to that subject.

A SIMPLE SWEEP GENERATOR

Basically, a sweep generator, or time-base generator, can consist of a capacitor that charges up through a resistance, building up a voltage between the junction of the resistor and capacitor (Fig. 3-1) and ground. When this voltage reaches a given level, a shorting device, switch S, discharges the capacitor as rapidly as possible. The result is a sawtooth waveform across the capacitor.

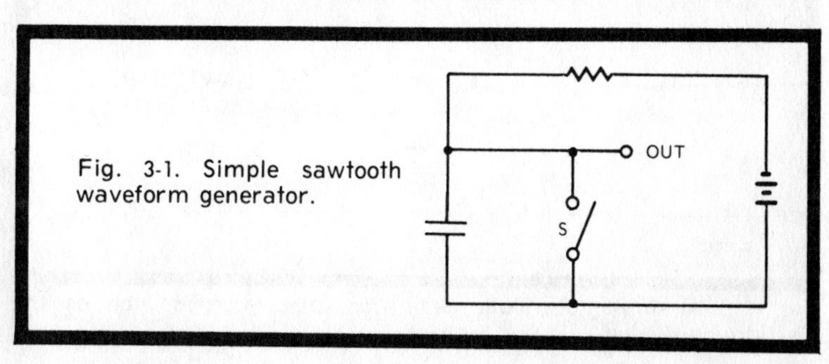

Fig. 3-1. Simple sawtooth waveform generator.

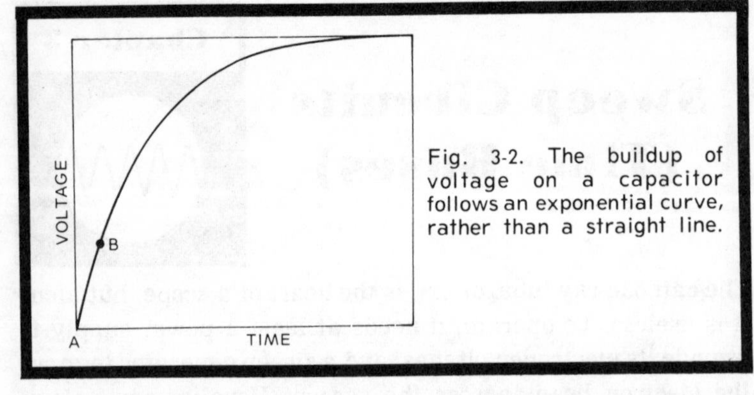

Fig. 3-2. The buildup of voltage on a capacitor follows an exponential curve, rather than a straight line.

But the buildup of voltage across a charging capacitor is not linear. That is, it is not proportional to time. In fact, the greater the voltage becomes, the slower its rate of increase (Fig. 3-2). The curve of such a voltage is said to be **exponential** rather than linear.

A sawtooth with a linear rise, or ramp, is required in an oscilloscope sweep circuit, and there are two main ways to obtain it from the circuit just described. One way is to limit the voltage built up on the capacitor to a fraction of the total available voltage. In Fig. 3-2 for example, only the small part of the charging curve between points A and B might be used. In this range, the charging curve is nearly linear, and from this bottom-most part of the curve a good sawtooth can be obtained. Because the voltage is limited for the sake of linearity in this method, a sweep amplifier following the sawtooth generator will be required.

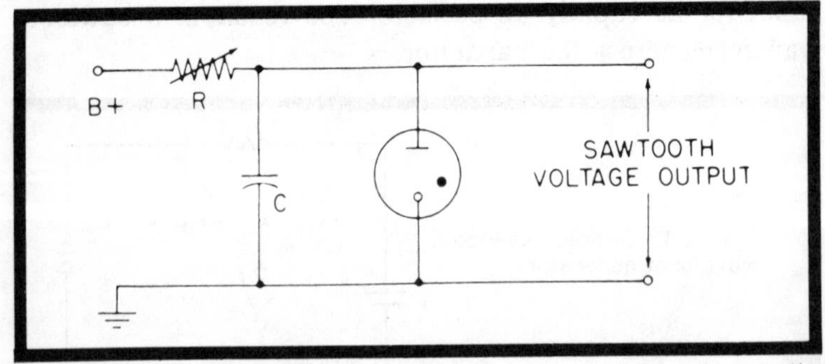

Fig. 3-3. A simple sawtooth generator using a neon tube as the discharging device.

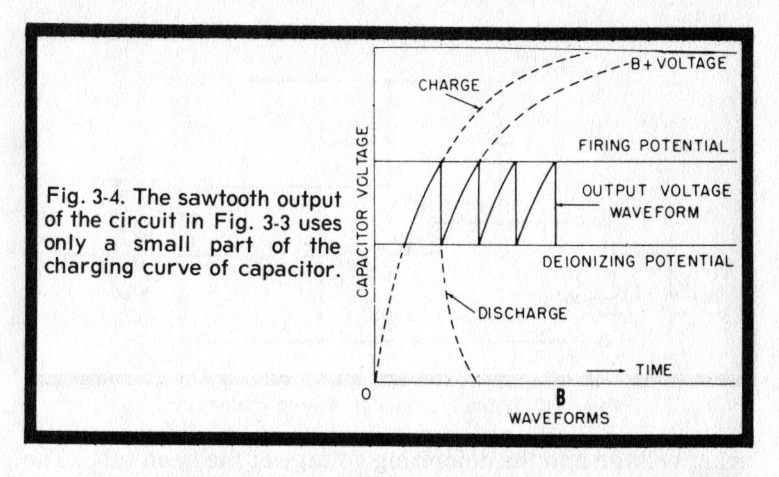

Fig. 3-4. The sawtooth output of the circuit in Fig. 3-3 uses only a small part of the charging curve of capacitor.

Another solution to the problem of linearity is to charge the capacitor at a constant rate, using a constant-current generator, which may be a transistor circuit. Sweep generators using this approach produce an output voltage of sufficient amplitude to directly drive a crt.

The very simple sawtooth generator in Fig. 3-3 is similar to the one in Fig. 3-1, only here a neon tube is used as the discharging device instead of a switch. In Fig. 3-3, the capacitor charges through the variable resistor until the voltage across the tube is sufficient to ionize the neon. When the ionization voltage is reached, the internal resistance of the tube suddenly drops from a high to a low value, and the capacitor is able to discharge through the tube. But when the voltage falls below the ionizing value of the neon tube, the high resistance of the tube is restored and the capacitor can no longer discharge through it. Thus, the capacitor starts to charge up all over again. This process is repeated over and over, and a repetitive series of identical sawtooth pulses result.

How fast the capacitor charges up depends partly on the value of the resistor, which is adjustable. If it is a high value, the capacitor will charge slowly. Conversely, if it is a low value, a large current can flow, quickly charging the capacitor. Thus, the variable resistor is able to control the frequency of the neon sawtooth generator. Fig. 3-4 shows how the sawtooth output voltage varies between the ionizing, or

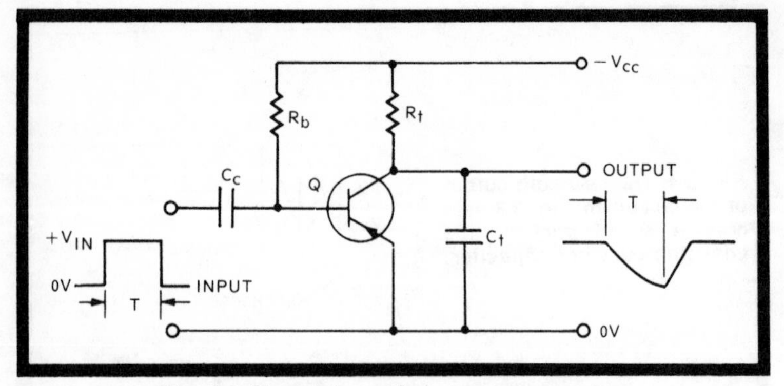

Fig. 3-5. Transistor-switch sweep generator.

firing voltage and the deionizing voltage of the neon tube. The total voltage is not reached or even approached, because the firing voltage is relatively low. By the same token, the voltage output never reaches zero either. The capacitor does not discharge completely, because when the voltage across it reaches the deionizing potential of the neon tube, the discharging stops. As Fig. 3-4 shows, over the range of voltages to which the output is limited, the charging curve is fairly linear, and the output of the circuit is a fair approximation of a sawtooth wave.

The circuit of Fig. 3-3, while it serves to illustrate the general principles underlying sweep generators, is not used in practice. Most modern oscilloscopes use transistor sweep generator circuits.

SIMPLE TRANSISTOR SWEEP GENERATOR

The circuit in Fig. 3-5 uses a transistor as a simple switch to control an RC (resistor and capacitor) network. Here the transistor is biased so that in the absence of any input signal it is in saturation. What this means is that the current in the base resistor R_b holds the transistor fully switched on, and the collector voltage is close to zero volts. In this condition, the transistor is a virtual short, and capacitor C_t is discharged. If at this point a positive pulse of duration t is applied to the input by way of the large coupling capacitor, C_c, the transistor base is driven positive and transistor Q is cut off for the duration of

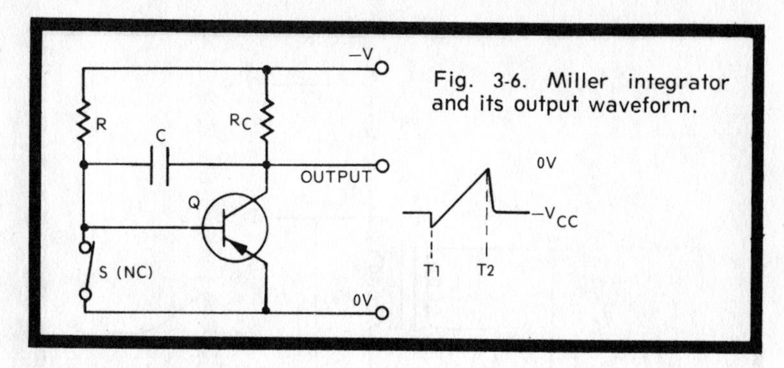

Fig. 3-6. Miller integrator and its output waveform.

the pulse. With Q cut off, capacitor C_t charges up through R_t, its upper plate acquiring a negative charge over the time t, as shown by the output waveform in Fig. 3-5. At the end of the positive input pulse, the base is no longer positive, and the transistor is driven into saturation again through R_b. The transistor then rapidly shorts out and discharges C_t, bringing the collector voltage quickly back to zero. The output voltage during interval t is exponential in form, but if pulse duration t is small compared to the time constant R_tC_t, the sweep output will approximate a sawtooth. That is because only a small part of the charging curve of timing capacitor C_t is used. The fact that the sawtooth is inverted in Fig. 3-5 is, of course, no problem. If the ungrounded terminal of C_t is connected to the left hand horizontal deflection plate, the electron beam will be swept from left to right as the voltage at that terminal becomes progressively negative during time t.

MILLER INTEGRATOR CIRCUIT

An even more commonly used sweep generator or time base is the Miller integrator circuit of Fig. 3-6. Transistor Q in this circuit is biased so that when switch S is closed the transistor is held cut off, the base being shorted to ground so that the transistor is deprived of base drive current. Thus the collector is at approximately the negative supply voltage $-V_{cc}$ and the base is near zero. C_t is thus charged, its right hand plate being negative. If the switch is opened at t1 and closed at t2, the output voltage takes the waveform shown. Immediately after the switch is opened, the base goes a few

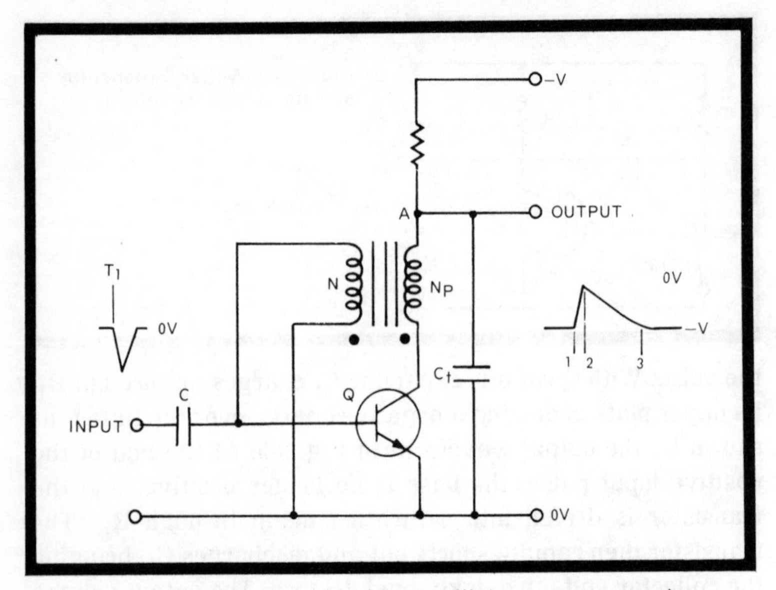

Fig. 3-7. Transistorized blocking oscillator sweep generator.

tenths of a volt negative as input current is switched into the base via R_t. This small negative blip is transmitted by C_t to the output, and is characteristic of Miller integrators.

Once switch S (which would be a transistor acting as a switch) is opened, C_t begins to charge through R_t with a time constant approximately proportional to C_tR_t. But the ramp waveform corresponding to this is **linear** instead of exponential. The reason is that due to feedback from output to input via C_t the effective capacitance of C_t is multiplied by the voltage gain of the transistor used. C_t acts like a much larger capacitor having a much longer charging curve. The ramp of the sawtooth output is but a small portion of this long charging curve, and thus is very linear. The capacitance multiplication by the amplifier gain, by the way, is called the Miller effect, and this is where the name **Miller Integrator** comes from.

BLOCKING OSCILLATOR SWEEP GENERATOR

In the blocking oscillator sweep generator of Fig. 3-7, the transistor base is normally placed at ground by the transformer secondary winding, so that the transistor is cut off.

The transistor collector and the output are both virtually at the negative supply potential. Capacitor C_t is charged up to $-V_{cc}$. If at this time a negative trigger pulse is applied to the transistor base via C_b, the base-emitter diode will go into forward conduction and the collector current will commence. This will cause the collector voltage to start moving from $-V_{cc}$ toward zero. This positive-going change in the collector voltage is phase-inverted by the transformer and applied to the base as a negative-going voltage. The phasing dots on the transformer symbol should clarify this. The negative-going voltage supplied by the transformer adds to the negative trigger voltage, increasing the collector current. This circular process quickly causes the transistor to go into saturation, reducing the voltage at point A to near ground potential, or zero volts. The capacitor C_t is fully discharged.

The current through the transistor builds up until it is limited by some circuit parameter. When the current stops increasing, the feedback voltage to the base disappears, and the transistor starts to cut off. This causes the collector current to drop, and the collector to move back toward the negative supply voltage again. The negative-going collector voltage is reversed and fed back to the base as a positive-going voltage, which further reduces the transistor current. The process "snowballs" so that the transistor is quickly cut off. Now the capacitor begins to charge up negatively through R_t towards $-V_{cc}$, as shown by the second part of the waveform in Fig. 3-7. Finally, the transistor returns to $-V_{cc}$ at t3. Between times t2 and t3, the output voltage provides a nearly linear ramp suitable for sweep generator use. The sweep however, is **exponential**, and to make it more linear, the blocking oscillator may be combined with other circuits. We will give an example of this shortly.

BOOTSTRAP SWEEP GENERATOR

The bootstrap circuit of Fig. 3-8 is basically an emitter-follower with feedback from the emitter via C_t to the junction of resistors R_f and R_t feeding current to the base. Note the timing capacitor between base and ground. The capacitor is

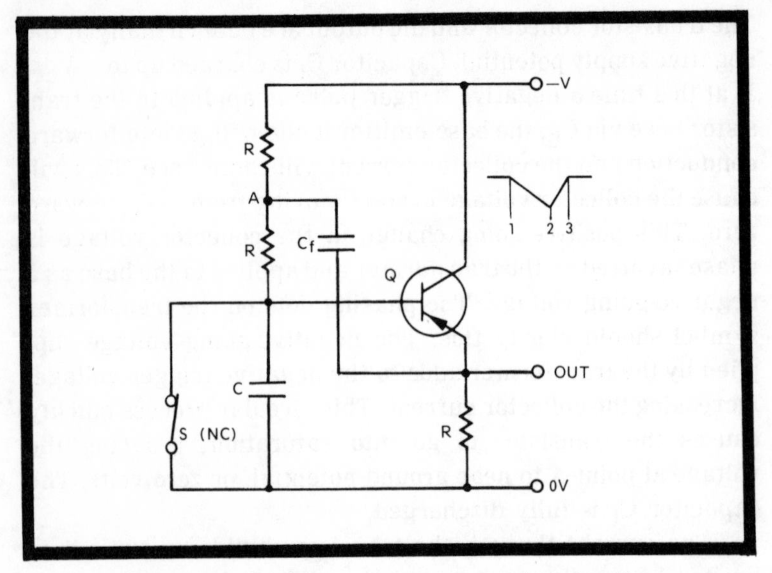

Fig. 3-8. Bootstrap sweep generator circuit.

normally shorted by the switch in Fig. 3-8. (In practice, a transistor circuit would replace the switch.) The base being at dc ground, the transistor is cut off and its emitter voltage is zero. If at time t1 the switch is opened, the output waveform takes the shape shown. C_t starts to charge up through R_f and R_t. As the voltage at the base of the transistor begins to go negative, so does the emitter output voltage, there being no phase reversal between input and output in an emitter-follower. This emitter voltage change is applied to the top end of resistor R_t through feedback capacitor C_f, making point A more negative. Since point A and the transistor base are both made more negative by the same amount, the voltage across R_t is virtually constant. If the voltage is constant, then the current must be constant. Thus, a constant charging current charges C_t between times t1 and t2. The output voltage follows the base voltage, and so we have the linear ramp shown in the figure.

At t2, switch S is closed again. This causes the transistor to return to the cutoff state and capacitor C_t to rapidly discharge, producing the retrace part of the waveform between t2 and t3.

A COMPLETE, PRACTICAL SWEEP GENERATOR

We said in our discussion of the bootstrap generator that the switch shown at its input in Fig. 3-8 would be replaced by a transistor circuit in actual practice. The transistor switch circuit could be a blocking oscillator (discussed earlier). Supplying voltage pulses with an oscillator may be likened to switching an ordinary electrical switch off and on very rapidly. (Of course it is quite impossible to operate an electrical switch at the frequency of which an oscillator is capable.)

The circuit in Fig. 3-9 is a **free-running** sweep generator. To say that a sweep generator is free-running means that it has a large degree of self-determination. The circuit shown has its own oscillator for determining the frequency of the sweep. Transistor Q1 is employed in a blocking oscillator circuit operating at about 10 kHz. Certain modifications have been made to the blocking oscillator to make it free-running, or **astable**, so that it requires no triggering pulse at its base for each operation. These modifications to the blocking oscillator previously described include the addition of R and R_b. Thus, the blocking oscillator supplies its own input pulses.

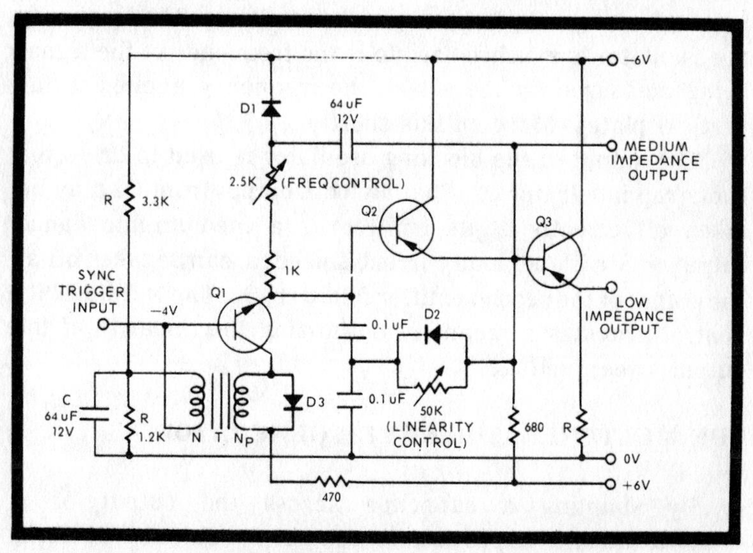

Fig. 3-9. A complete free-running sweep generator.

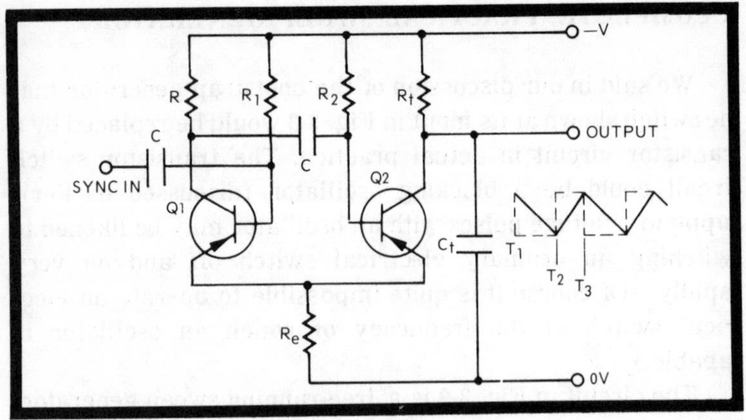

Fig. 3-10. A sweep generator consisting of a multivibrator with a timing capacitor across its output.

Although the free-running frequency of the oscillator is set at about 10 kHz by its fixed components, the frequency can be varied over a certain range by the variable resistor **frequency control**. This rheostat would be adjusted by a knob on the front panel of the scope.

The sync trigger input provides a means of synchronizing the blocking oscillator to an external control frequency, provided the control frequency is slightly higher than the preset, free-running frequency of the oscillator. The frequency the oscillator is synchronized to is the frequency of the signal being measured by the scope, the frequency applied to the vertical plates. More on this shortly.

The output of the blocking oscillator is used to drive the bootstrap integrator Q2. The sawtooth output from Q2 may be taken off directly at its emitter as a medium-impedance output or, if a higher load current is needed, can be taken off at the output of the second emitter follower, Q3. The 50K linearity control provides a means of optimizing the linearity of the output sweep voltage.

THE MULTIVIBRATOR SWEEP GENERATOR

By shunting a capacitor across the output of a multivibrator, it is possible to convert its output waveform from a rectangular wave to a sawtooth. In this manner, the

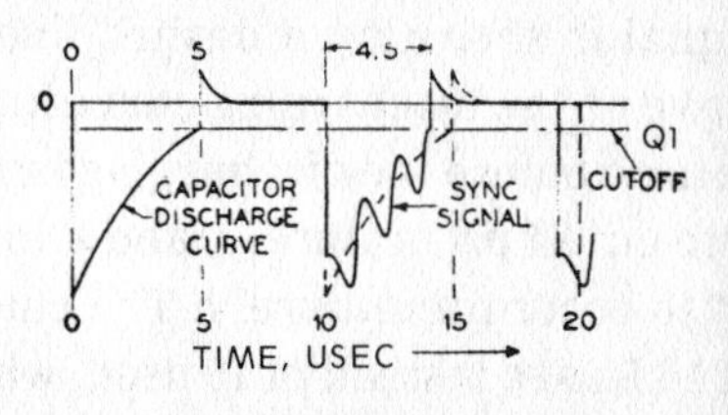

Fig. 3-11. The sine wave sync signal riding on the discharge curve causes transistor Q1 to be cut off prematurely.

multivibrator can be used as a sweep generator. In the example in Fig. 3-10, an emitter-coupled astable multivibrator is set up with a timing capacitor C_t connected between the Q2 collector and ground. Without C_t in the circuit, the multivibrator output would assume the rectangular form shown by the dotted lines. With capacitor C_t, the output waveform is a sawtooth.

From time t1 to time t2, Q2 is cut off. During this interval, C_t charges negatively through R_t with a time constant C_tR_t. From t2 to t3, when Q2 is switched on, C_t discharges through resistor R_e. Since a much smaller resistor is used for R_e than for R_t, the capacitor discharges more quickly than it charges. Thus, the retrace time t3-t2 is much shorter than the forward sweep time t2-t1.

In order for the display on an oscilloscope to stand still instead of rolling horizontally, it is necessary to synchronize the sweep generator and the signal applied to the vertical deflection system. To do this, some of the input signal may be applied to the sweep generator.

To synchronize the generator in Fig. 3-10 to an input signal, a sample of the input signal might be applied to the base of Q1 by means of the sync input shown. Suppose the waveform of the voltage of Q1 is as shown in Fig. 3-11. With no sync signal, the waveform might look as shown between 0 and 5 usec. During this time, when the voltage is below the value represented by the horizontal dashed line, Q1 is cut off, Q2 is switched on, and a forward sweep is occurring. Now imagine that at 10 usec and for all times thereafter a synchronizing voltage is applied to the base. Observe how the RC discharging curve is altered in the approximately 5 usec interval after 10 sec. The curve has a sine wave superimposed on it. The path the curve would have taken in the absence of a sync signal is shown by a dashed line. Observe how the sine sync wave on the discharging curve hits the cutoff voltage at a point in time before the discharging curve does. This causes Q1 to go into cutoff prematurely, and a forward sweep at the output of Q2 to occur prematurely. The cutoff of Q1 and the sweep occur at 14.5 usec instead of 15 usec, when they would have occurred without sync. Instead of taking 5 usec to reach the cutoff voltage, the base only took 4.5 usec. Similarly, on the next and all succeeding cycles the base reaches the cutoff level 4.5 usec after the capacitor begins to discharge.

In this manner the multivibrator output voltage is synced with the sine-wave voltage applied to the base of Q1. The sync signal causes the free-running multivibrator frequency to increase slightly when it is locked in with the frequency of the sync signal, since the cycles now take slightly less time than before.

Multivibrator sweep generators can be synchronized by square waves as well as sine waves, and quite commonly are. Now let's see just why synchronization is necessary.

SYNCHRONIZATION

The frequency of a time-base or sweep generator is mainly determined by an RC network. There are, however, other factors affecting the frequency of the generator. Variations in supply voltages and temperature effects contribute to temperature instability. Even the RC network is subject to variation under certain conditions. Thus, the same setting of the frequency control may produce different frequencies at different times. Under these circumstances, an oscillator is said to be free-running. The amplitude as well as the frequency of the sawtooth may vary slightly from cycle to cycle. This is intolerable in a scope, because even a minor variation in the sweep frequency will cause the display to drift across the face of the crt, making observation and measurement tough, and photography impossible. That is, frequency variation in the sweep generator of a scope will cause the display to roll across the screen the way a picture rolls upward or downward on a TV screen if the vertical oscillator of the TV drifts. For proper operation of either a TV or scope, some means of controlling the sweep frequency is necessary. That means is synchronization, shortened to **sync**. By sync, the frequency of the sweep oscillator is controlled by a source outside its own circuit. If an accurately maintained frequency is applied to the input of the sweep generator, this standard frequency will start the sweep at the same point in each standard cycle, making all cycles of the generator output the same in size and duration. Thus, the proper frequency will be maintained at all times.

It is not really necessary that the sync signal or pulse be absolutely constant in frequency. The main thing is to have the sweep generator follow the sync signal as closely as possible. If the frequency of the incoming signal varies, and the frequency of the sweep generator varies right along with it, the display will stand still rather than drift.

RELATIONSHIP BETWEEN SWEEP AND SIGNAL FREQUENCIES

It is usually desirable to view more than one cycle of a signal on a scope at the same time. This is because part of the signal is lost during retrace at the extreme left and extreme right of the screen. Suppose the frequency of the signal to be

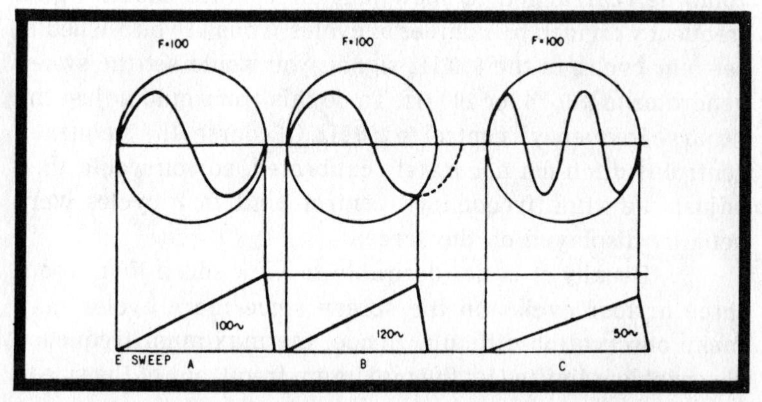

Fig. 3-12. The relationship between the signal and the sweep voltage when (a) the sweep time equals the period of one cycle, (b) the sweep time is less than the period of a cycle, and (c) when the sweep time equals the period of two cycles of the signal.

studied is 800 Hz. It takes one eight-hundredth of a second for a cycle of the signal to occur. Now suppose the electron beam moves across the screen once in that time in response to the ramp portion of a sawtooth cycle. One cycle of the signal will be displayed, as shown in Fig. 3-12(a). Notice that the first part of the next sine cycle, indicated by the dotted line, will be lost during retrace. Thus, instead of appearing on the left side of the screen as the forepart of the first cycle did, the forepart of the next and all succeeding cycles will be lost. So instead of the wave in Fig. 3-12(a), you would see an incomplete wave.

If the beam were to move across the screen faster, that is, if it made a left to right excursion across the screen **before** the sine wave was completed, the part of a cycle represented by the dotted line in Fig. 3-12(b) would be lost. Much less than one full cycle would be displayed.

On the other hand, if more than one cycle of the signal occurred during one left to right excursion of the beam, then more than one cycle of the signal would be displayed. If two cycles of signal occurred during one sweep, two cycles would be displayed on the screen, as in Fig. 3-12(c). Thus you can see that the ratio between signal frequency and sweep frequency is the number of cycles displayed. An equation, showing number of cycles equal to signal frequency over sweep rate could be rearranged to show that sweep rate equals signal frequency divided by number of cycles. Thus, if you wished to see four cycles of the 800 Hz signal, you would set the sweep generator to 800/4, or 200 Hz. To do this you would adjust the "coarse frequency" control to 200 Hz. Of course, the frequency control is often not accurately calibrated, so you would then adjust the "fine frequency" control until four cycles were actually displayed on the screen.

Usually it is not desirable to view more than about three or four cycles on the screen since more cycles may make observation difficult. Hence, the maximum frequency observable is limited by the maximum frequency of the sweep generator. The maximum observable signal frequency is equal to the maximum sweep frequency times about 4. If the maximum sweep frequency is 100 kHz and the greatest number of cycles it is permissible to have in the display is four, then the maximum signal frequency is 4 times 100 kHz, or 400 kHz. Of course, higher signal frequencies are observable if more display cycles can be tolerated. Note that in many cases the maximum frequency observable with a given scope is limited by considerations other than the time base, or sweep generator.

The fewer the number of cycles to be displayed, the higher the sweep frequency required. Often it is desired to view a small part of a wave and the sweep generator cannot sweep the screen fast enough to permit this. One way to get around the problem is to display a single cycle and then increase the horizontal gain beyond the width of the screen, so that a small part of the signal wave occupies the whole screen. Advanced scopes (discussed later) using calibrated sweep delay techniques provide a more desirable, albeit expensive, way of viewing parts of cycles.

TRIGGERED SWEEP

As you have seen, it is necessary to sync both the signal being observed with a scope and the sweep generator. This can be accomplished fairly simply, as explained earlier, when the input signal is a regularly recurring one, such as a series of sine waves. But what about when the input signal consists of pulses that occur randomly rather than regularly at some specific frequency? No matter what frequency we set for the sweep generator, its frequency cannot be synced to the frequency of the pulses, since they have no set frequency. To observe such pulses, triggered sweep is required. This is a feature of all lab-type or engineering scopes.

A triggered sweep is obtained by using a one-shot sweep oscillator, that is, one that operates once and then stops. A small portion of the input signal itself is applied to a trigger circuit. When the input signal reaches a predetermined threshold voltage, the trigger, which may be a type of multivibrator, is tripped. A very narrow voltage spike is created by the trigger circuit and applied to the one-shot

Fig. 3-13. The EICO TR-410 laboratory quality, triggered-sweep is an outstanding buy. (Courtesy: EICO Electronic Instrument Co., Inc.)

sweep oscillator, initiating its operation and the horizontal sweep. When the sweep is completed, the sweep oscillator returns to rest, awaiting the next signal cycle or pulse. This method of synchronization is good for continuous signals as well as for random pulses. Since the sweep oscillator does not operate continuously, it has no frequency. It merely sweeps the electron beam across the screen at a certain speed, which is adjustable. Rather than adjust the frequency of such a generator, we adjust the **time** it takes to move the beam across the one division on the screen. Thus, the triggered-sweep circuit is a **time** base in the truest sense. Because triggered-sweep scopes are becoming so common, "time base" is rapidly replacing "sweep generator" as the term describing the horizontal deflection generator is an oscilloscope. "Time base" is used exclusively when discussing lab-type scopes. The EICO TR-410 of Fig. 3-13 is an example of a modern, triggered-sweep scope.

CRT UNBLANKING

Sawtooth waves generated by practical circuits are not perfect. Not only are their ramps always at least slightly nonlinear, their flyback to zero volts is not instantaneous. Thus, the electron beam requires a definite time to return to the left side of the screen after a sweep. It is usually desirable to remove the electron beam from the screen during this time. The reason why can be illustrated by a simple example. As you write across a sheet of paper from left to right, periodically you reach the right edge of the paper and have to return to the left to start the next line. Rather than just drag the pencil back to the left, you remove it from the paper and return it to the left side of the paper to start the next line. It is similar with the electron beam of a crt. Writing across the screen from left to right, the beam periodically reaches the right edge of the screen and has to return to the left. Just as retrace lines from right to left on a handwritten page already containing text would be confusing, retrace lines on a crt screen are confusing. So the beam is removed from the screen before it is returned to the left. Sometimes the beam may be deflected off-screen during retrace, but in most oscilloscopes, the beam is simply cut off while the voltage on the horizontal deflection plates is set to the polarity and value needed to put the electron beam at the left side of the screen when it is turned on again. Cutting off the electron beam during flyback (retrace) is called **blanking**.

There are two main ways in which blanking is accomplished. One way is to apply a strong negative pulse to the grid of the crt during retrace. Such pulses take the form of square waves and are to be found somewhere in most sweep generators. Often the pulses are amplified by a blanking amplifier before being applied to the crt grid.

In some scopes a strong negative bias is applied to the grid of the crt, biasing it into cutoff or blanking. Just as a sweep begins, a positive square wave is applied to the grid, overcoming the cutoff bias and turning the beam on. Since this type of pulse turns the beam **on** instead of off, it is called an **unblanking** pulse. This blanking method we are talking about now involves more complicated circuitry than the other method, and is usually reserved for high-performance, triggered-sweep scopes.

Retrace is not much of a problem at low frequencies, but at high sweep frequencies, where the retrace time is no longer negligible compared to the sweep time, it can be a problem in a scope without blanking. On the other hand, blanking is not always desirable, since it renders invisible a portion of the pattern of the signal being tested. For this reason, some scopes have an **optional blanking** feature, permitting operation without blanking when desired.

Oscilloscope Circuitry

In the preceding chapter we considered the specialized circuitry found in scopes. In this chapter we shall consider the other circuits that go together to make up a scope. First let's see how these circuits are organized into a system.

THE SCOPE AS A SYSTEM

Fig. 4-1 is a block diagram of a typical triggered-sweep scope, with the power supplies omitted. The timing diagram in the center of the block diagram shows the time relationship between the waveforms found at various points in the oscilloscope. Refer to Fig. 4-1 as we follow these relationships.

The waveform to be observed, A, is applied to the vertical amplifier input. A calibrated volts-per-division control adjusts the gain of the vertical amplifier. In some scopes, this control is called **vertical gain**. The push-pull output of this circuit, B and C, is fed through a delay line to the vertical deflection plates of the crt. The reason for using the delay line will be shown shortly.

The sawtooth wave, E, developed by the time base (sweep generator) is used to deflect the electron beam in the crt across the screen from left to right. The rising or run-up portion of this wave is linear, rising through the same number of volts during each unit of time. By means of a calibrated time-per-division control, the rate of rise is set to the desired value. The output of the time base is applied to a time-base amplifier, which includes a phase inverter. The output of the time-base, or horizontal, amplifier consists of two sawtooth waveforms: one positive-going like the input, and one negative-going as a result of phase inversion. The positive-

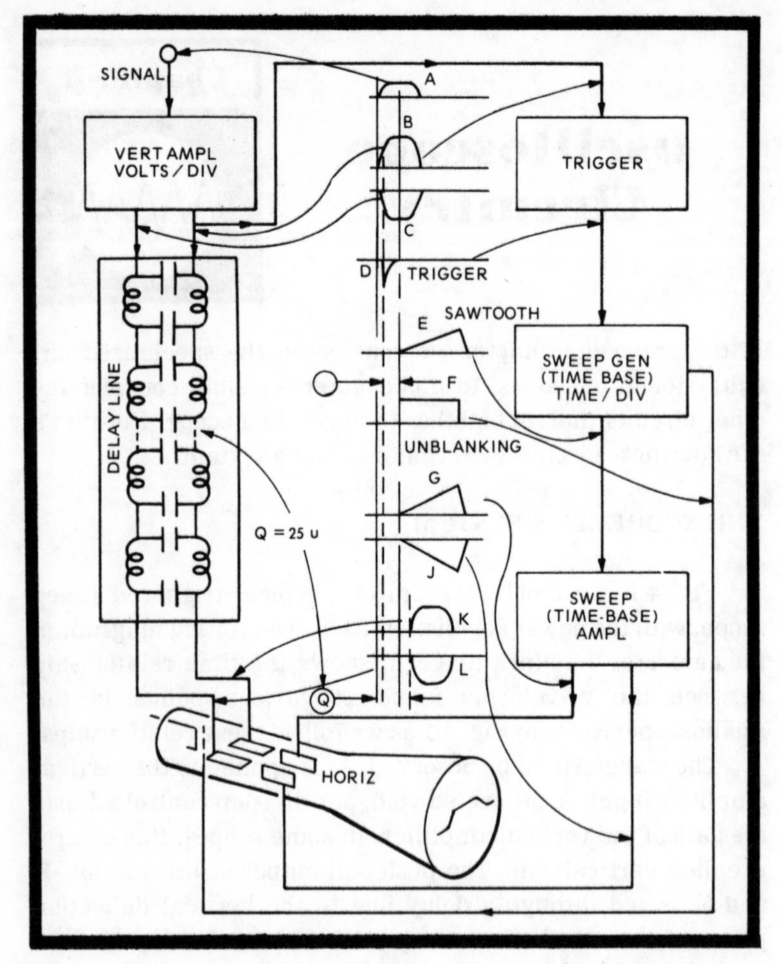

Fig. 4-1. Functional diagram and timing diagram of a triggered-sweep scope.

going sawtooth, G, is applied to the right-hand horizontal deflection plate, and the negative-going one, J, is applied to the left-hand plate. Consequently, the beam is swept horizontally to the right through a given number of divisions during each unit of time. The rate at which the beam is swept across the screen, as we said, is determined by the setting of a time-per-division control. This control may be compared to the throttle of your car, which determines the time you take to go each mile. The time-per-division control sets the speed of the beam's travel across the screen—although "time-per-division" is actually the reciprocal of the speed (1 divided by speed), just as hours per mile is the reciprocal of a car's speed. In scopes with a free-running sweep generator—that is, in scopes without triggered sweep—a **horizontal frequency** control is used instead of the time-per-division control. Of course, the higher the horizontal frequency, the faster the sweep of the electron beam.

To maintain a steady, drift-free display on the screen, each horizontal sweep must start at the same time on the input waveform applied to the vertical plates. To accomplish this, a sample of the input waveform is applied to a trigger circuit, which produces a negative voltage spike, D, at some selected point on the displayed (input) waveform. The run-up part of the time-base sawtooth is initiated by this triggering spike. Triggering occurs at the start of the horizontal sweep at the left side of the screen.

A rectangular unblanking pulse, F, is applied to the crt grid and corresponds exactly in time to the run-up portion (ramp) of the sweep sawtooth. This turns the beam on during the run-up time so that it can create a trace on the crt face. At all other times, the beam is cut off by a negative bias voltage on the grid. Cathode ray tubes in high-performance scopes use large accelerating voltages, which accelerate the electrons in the beam to high velocities. These high-velocity electrons might damage the scope screen if the beam weren't cut off during the time it is not being swept across the screen.

Now let's take a closer look at the timing in the scope system. In Fig. 4-1, the leading (left-hand) edge of the waveform being observed is used to kick on the trigger circuit. It takes a certain amount of time, about 0.15 usec (P), for the triggering and unblanking operations. Thus the sweep of the electron beam does not occur exactly simultaneously with the leading edge of the input waveform. To allow us to see the leading edge, a delay, Q, of about 0.25 usec is introduced into the vertical deflection channel, after the point in the channel where a portion of the vertical signal is taken off and fed to the trigger circuit. Thus, the application of the vertical signal (K and L) to the vertical deflection plates is held off until the triggering and unblanking operations have been completed

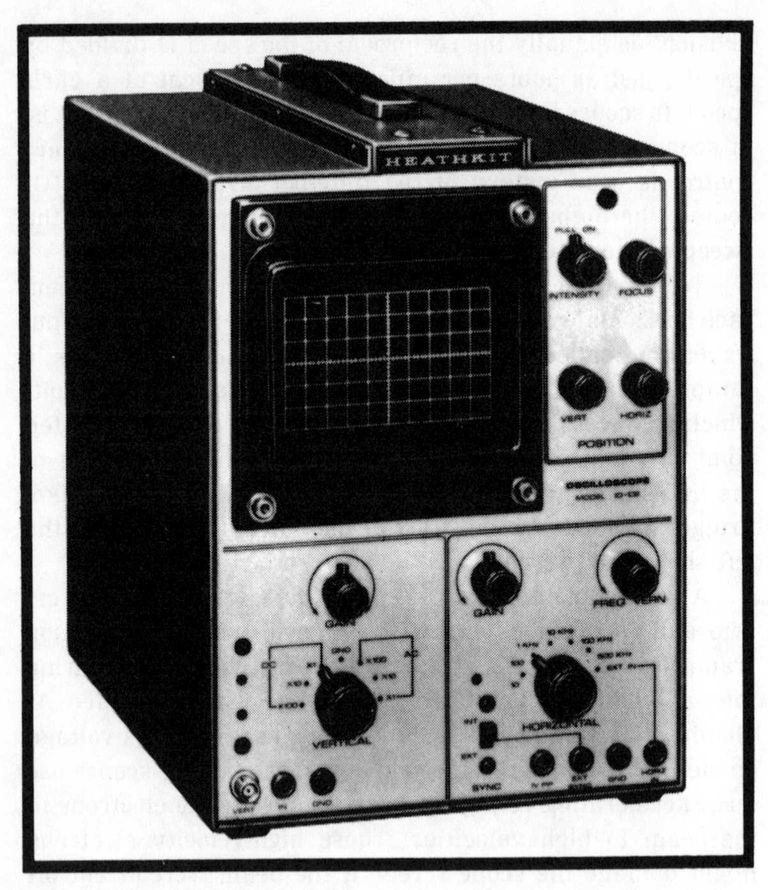

Fig. 4-2. The Heath IO-102, whose circuitry is explained in the text, is an all-solid-state service-type scope. (Courtesy Heath Co.)

and the beam is ready to sweep the screen. This way, even though the leading edge of the input signal is used to trigger the sweep, the leading edge may still be observed. If there were no vertical delay, only the part of signal represented by the right half of waveform B could be observed.

TYPICAL SCOPE

Heathkit's IO-102 5-inch scope in Fig. 4-2 is a general purpose TV service type scope. Except for the crt, the IO-102 is all-solid-state. The solid-state construction results in a weight of just 27 pounds and compact overall dimensions.

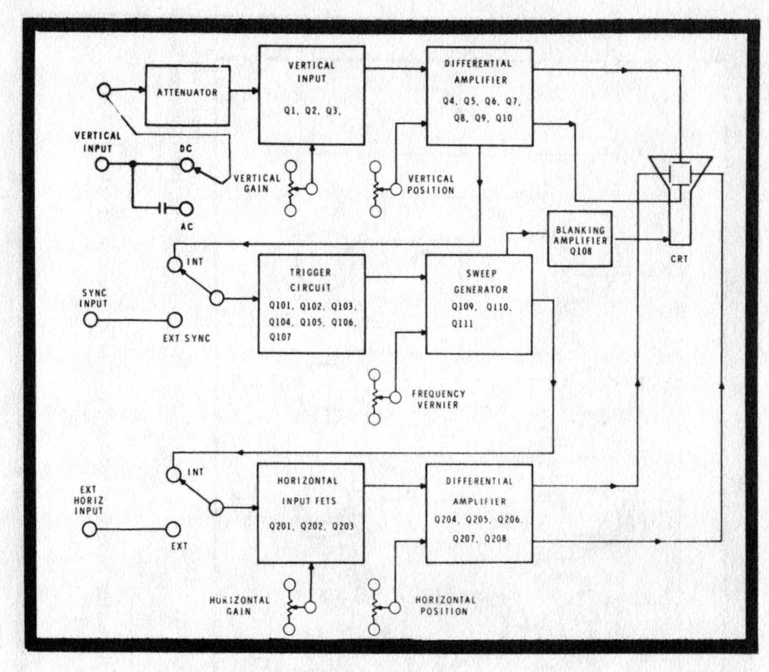

Fig. 4-3. Block diagram of the Heath IO-102 scope. (Courtesy Heath Co.)

Although the crt is a circular tube having a diameter of 5 inches, the actual viewing area is a 6 by 10 cm rectangle. The transparent **graticule**, mounted in front of the crt screen, is marked off in horizontal and vertical divisions of one centimeter. Note that there are six divisions vertically and ten divisions horizontally.

Follow along on Fig. 4-3 as we trace through the circuits of the IO-102 scope. Note that a trigger circuit is shown. This does **not** mean that this scope is a triggered-sweep scope. It isn't. It has a recurrent sweep adjustable between 10 Hz and 500 kHz. The trigger circuit generates a pulse that starts the sweep generator a bit earlier than it would start on its own. This pulse is generated in response to a sync signal, which may be taken off the vertical deflection channel or from some outside source. Even without the trigger circuit a sweep would be generated at least every 0.1 sec, since the lowest frequency of the sweep generator is 10 Hz and the generator runs continuously. In a triggered-sweep scope, no sweep would be produced until a signal in the vertical channel triggered it. If

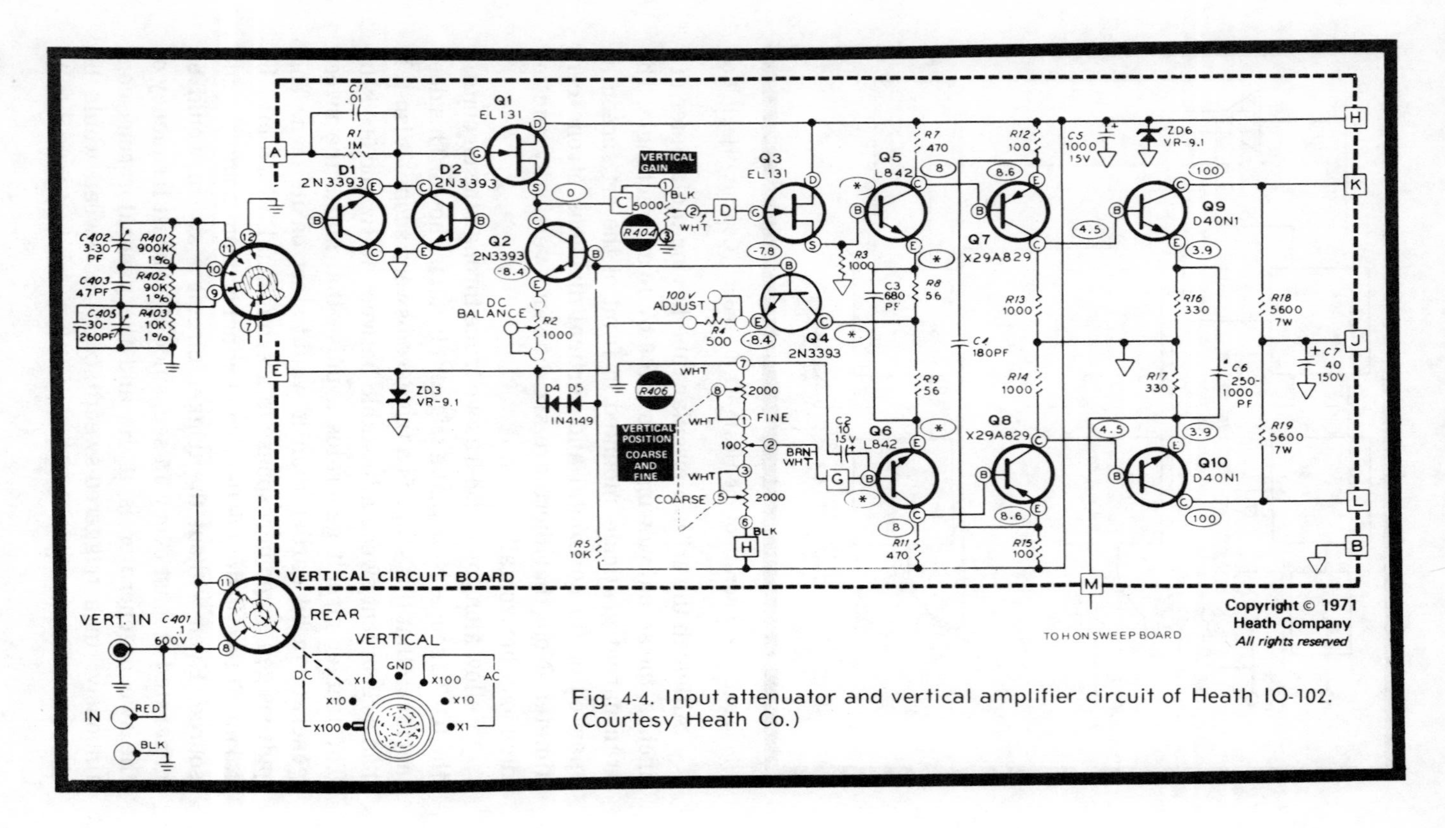

Fig. 4-4. Input attenuator and vertical amplifier circuit of Heath IO-102. (Courtesy Heath Co.)

the signal to be viewed on a triggered-sweep scope (not the IO-102) occurred only once every minute, then a horizontal sweep would occur only once every minute. In the IO-102, horizontal sweep occurs at the rate of at least 600 sweeps per minute. In this scope, the trigger circuit merely helps to lock the recurring sweep in with the signal to be examined.

Input Attenuator

A signal to be examined is applied across the vertical input attenuator, consisting of resistors R401 through R403 and capacitors C402 through C405 in Fig. 4-4. For weak signals, the vertical control is set to the x1 position, and the entire signal voltage, which appears across R401 to R403, is applied to the vertical amplifier at point A on the vertical circuit board. For stronger signals, the attenuator is set to the x10 position, and the part of the signal voltage appearing across R402 and R403 is applied to the vertical amplifier. This is 10 percent of the total signal input voltage applied at the vertical input terminals on the front panel of the scope. The signal is stepped down in this manner to prevent the beam from being driven off-screen during positive and negative signal peaks. For still stronger signals, the input must be stepped down even more. Thus, for these signals the attenuator is set to the x100 signal, so that only 1 percent of the signal is applied to the vertical amplifier. That is the part of the total signal appearing across R403.

The significance of the designations x1, x10, and x100 is as follows. If, when the vertical attenuator is set to the x1 position, a signal with a peak value of 1 volt causes a maximum deflection of one division, then when the attenuator is moved to the x10 position, it will take a 10-volt signal to cause a deflection of one division. And when the attenuator is set to the x100 position, it will take a 100-volt signal to produce a vertical deflection of one division. Thus, when the attenuator is set to x1, each vertical division represents 1 volt, when it is set to x10, each division is 10 volts, and when it is set to x100, each division represents 100 volts.

Whatever voltage causes a deflection of one division when the attenuator is set to the x1 position is determined by the setting of the **vertical gain** control. Unlike the attenuator, which is a switch, the vertical gain control is a continuously variable control. If the vertical gain is set so that 0.5 volt causes a one division deflection with the attenuator at x1, when the attenuator is moved to x10 it will take 5 volts to cause the same deflection. A one-division deflection represents 0.5 volt when the attenuator is at x1 and a voltage 10 times as great when the attenuator is at x10. Similarly, if the attenuator were moved to x100 **without changing the vertical gain control**, one division would represent a voltage 100 times as great as it did with the attenuator in the x1 position.

When the attenuator and vertical gain are set so that one division represents 1 volt, the scope is said to be calibrated for 1 volt per division. When these controls are set so one division represents 0.5 volt, the scope is calibrated for 0.5 volt per division. By means of the attenuator and vertical gain controls, the scope may be calibrated for just about any voltage value desired, and may be used as a voltmeter. The attenuator switch is analogous to the range switch of a voltmeter.

The capacitors in the attenuator, C402 through C405, provide compensation of the attenuator. If it were not for the capacitors, the frequency response of the attenuator would vary with the attenuator setting. Response is independent of frequency when the time constants of all the resistors and their associated capacitors are made equal, that is, when R401 times C402 equals R402 times C403 equals R403 x (C404 + C405). Notice that the capacitors are adjustable so that these time constants can be set equal.

Vertical Amplifiers in General

One of the characteristics of a crt is its **deflection factor**. In most crts, the deflection factor is about 60 volts. This means the luminous spot on the screen will be deflected 1 inch by a dc voltage of 60 volts or by an equivalent p-p ac voltage. The corresponding ac voltage is an rms voltage of half the product of 60 x 0.7, 21 volts. If the minimum height for a legible display is considered to be 1 inch, then rms voltages of less than 21 volts rms must be amplified for observation. This is the function of the vertical deflection amplifier. If the vertical amplifier has a gain of 1000, then rms voltages as small as 21 divided by 1000, or 21 millivolts, can be observed. The scope is said to have **a deflection sensitivity of 21 mV per inch**, more than adequate for most applications.

Another important characteristic of a vertical amplifier is its bandwidth—the range of frequencies over which it provides a constant gain. For low-frequency work, a bandwidth of 20 Hz to 100 kHz is adequate. For displaying video frequencies, an upper frequency limit of several megahertz is usually desired. Displaying digital signals requires a bandwidth of 0 Hz (dc) to about 50 MHz.

Heath IO-102 Vertical Amplifier

In the vertical amplifier of the Heath IO-102 in Fig. 4-4, a signal applied to the VERT IN connector on the front panel of the scope is coupled through the attenuator network to the vertical amplifier. (See also Fig. 4-3.) Capacitor C401 blocks the dc when the vertical switch is in the ac positions. The portion of the input signal selected by the attenuator is coupled through R1 and C1 to the gate of Q1. R1 protects Q1 from damage if a high potential is applied to the VERT IN connector while the vertical switch is in one of its lower range settings. Diodes D1 and D2 are actually transistors. They are connected to provide a zener action, which limits the signal applied to Q1 to plus and minus 9 volts. This helps to prevent excessive gate potential. Capacitor C1 improves the high-frequency response of the amplifier by forming a high-frequency bypass of R1.

Transistor Q1 is a field-effect transistor (FET) connected in the source-follower configuration. The use of a FET in the input minimizes loading of circuits under test, since the FET provides a high input impedance.

Transistor Q2 is a constant-current source for Q1. Diodes D4 and D5 each provide a 0.6 volt drop for a total of 1.2 volts, and hold the base of Q2 at a constant voltage. As the circuit of

Q2 is an emitter follower and the emitter voltage is dependent on the base voltage, the emitter voltage will also remain constant. This constant emitter voltage is across emitter resistor R2. Thus, the current through R2 is constant. Dc balance control R2 is adjusted so that the source voltage of Q1 is zero when an input signal is not present. A signal applied to the gate of Q1 will cause voltage changes at the source, but not current changes. These voltage changes are applied across the gain control, R404, and part of this signal is applied to the gate of source-follower Q3.

Transistor Q4 is used as a constant-current source for Q5 and Q6. Since the emitters of both Q5 and Q6 are connected to this constant-current source, the current source acts as a common emitter resistor and sets the operating point for Q5 and Q6.

The output of source-follower Q3 is amplified by Q5. A portion of the signal applied to the base of Q5 appears at its emitter. Since Q5 and Q6 share the same emitter resistance, the signal present at the emitter of Q5 is also present at the emitter of Q6.

Transistor Q6 is a common-base amplifier whose base is held at a constant voltage by the vertical position control, R406. This control positions the display on the crt vertically by applying a dc voltage to the base of Q6, thus causing an intentional imbalance in the vertical amplifier. As the collector output voltage of Q5 **decreases**, the emitter voltage of Q5 **increases**; an increased emitter voltage at Q6 reduces its forward bias and **increases** its collector voltage. The signal at the output of Q6 is 180 deg out of phase with the signal at the collector of Q5. Thus, Q5 and Q6 create the push-pull type of output desired for driving the vertical deflection plates of the crt. Capacitor C3 is an emitter bypass capacitor that boosts the gain of the push-pull amplifier at high frequencies. This results in an extension of the upper frequency limit to 5 MHz. Emitter resistors R8 and R9 establish the dc gain of the vertical amplifier.

Transistors Q7 and Q8, in addition to providing some gain, isolate transistors Q5 and Q6 from output transistors Q9 and Q10. Q7 and Q8 can be seen to be connected as common-emitter amplifiers. Capacitor C4 couples the emitters of Q7 and Q8 together for high frequencies, improving the high-frequency response.

Q9 and Q10 constitute a push-pull amplifier that is used to drive the vertical deflection plates. There are two reasons for using a push-pull arrangement for driving the plates. One reason is that the use of a single-ended amplifier for this purpose results in **trapezoidal distortion**. In this type of distortion, the plates nearest the electron gun (either the vertical or the horizontal plates) influence the signal on the other plates. If the vertical plates are nearest the gun, the horizontal plates may be affected in such a way that nonlinear sweep occurs. Another type of distortion is avoided by push-pull operation. That is the distortion due to overloading the output stage. With push-pull operation, each transistor has to provide only half the total deflection voltage, so that overload distortion is prevented or reduced.

IO-102 Sweep Generator

The sync switch selects either a portion of the amplified vertical input signal, or a signal applied to the external sync connector. When the sync switch is in the INT (internal) position, a portion of the signal on the emitter of Q10 is coupled via point M on the vertical circuit board, point H on the sweep circuit board, and capacitor C102 to the gate of transistor Q101. See Fig. 4-5. When the sync switch is in the ext (external) position, an external sync signal may be coupled to the gate of FET Q101, a source follower. D102 and D103 are transistors connected as diodes to protect the input of Q101 from excessively high external sync signals. The diodes do this by clipping at plus and minus 9 volts. Q102, a constant-current source for Q101, is adjusted by sync level control R103 to provide proper biasing for the sync circuits. This is to insure that even a small sync signal can synchronize the sweep generator.

Next the signal is amplified by Q103 and Q104 and applied to the Schmitt trigger circuit, consisting of Q105 and Q106. The Schmitt trigger is a bistable multivibrator with feedback; it

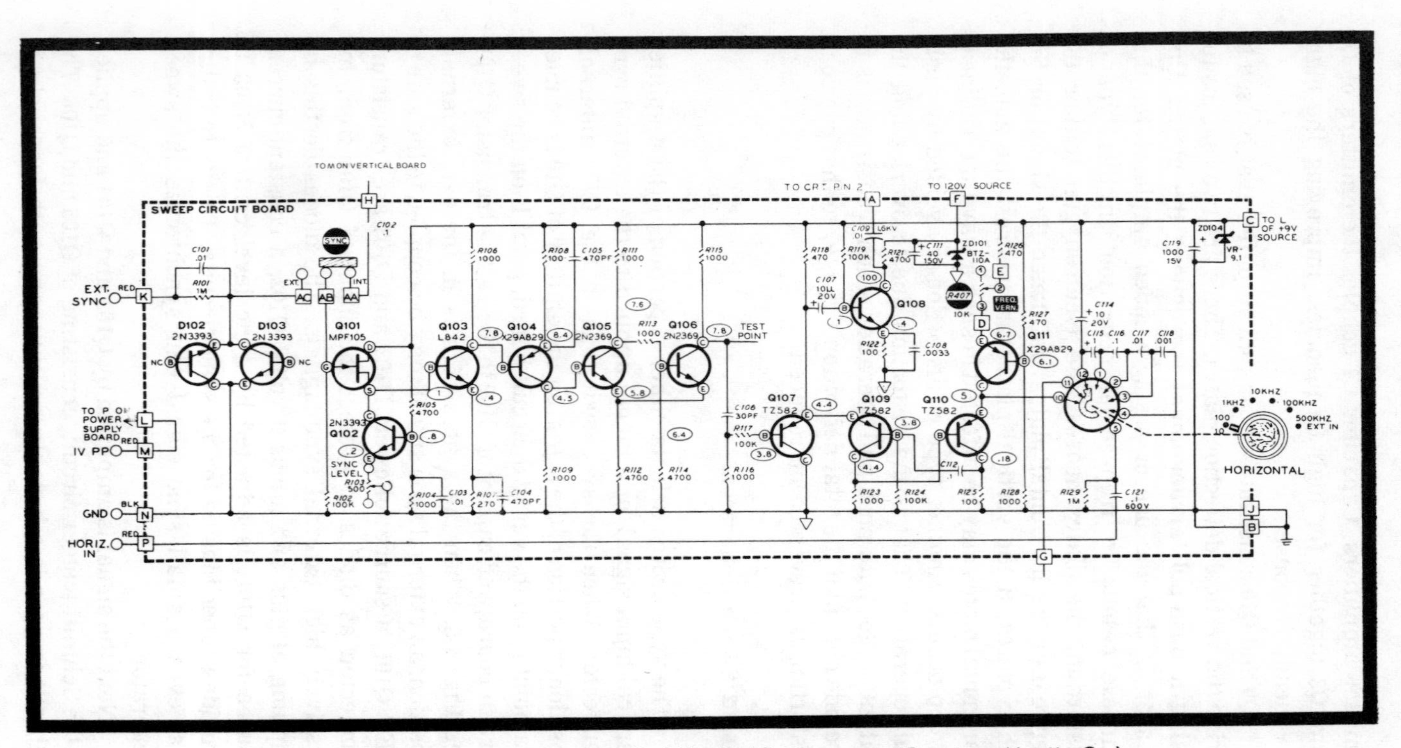

Fig. 4-5. Sweep generator circuit of Heath IO-102 scope. (Courtesy Heath Co.)

produces a rectangular pulse at its output each time it is triggered and reset.

Transistors Q109 and Q110 form an astable (free-running) multivibrator. When Q110 is on and Q109 is cut off, one or more of the timing capacitors (C114 through C118), as determined by the setting of the horizontal (sweep frequency) switch, are charged through Q110. As the voltage at the emitter approaches the voltage at its base due to the increasing voltage on the charging capacitor, Q110 will cut off and drive Q109 into conduction. With Q110 cut off, constant-current source Q111 can supply current to the timing capacitors. This current is of such polarity as to discharge the timing capacitors. The setting of the frequency vernier (fine tuning) control determines the current flowing through Q111, and thus the discharge current and time of the timing capacitors. As the timing capacitors discharge, a positive-going ramp voltage is generated and fed to the horizontal amplifier. The frequency of the sweep is determined in stages by the setting of the horizontal control, and continuously between stages (decades) by the frequency vernier control.

Inasmuch as Q107 and Q109 have a common emitter resistor, a signal applied to the base of Q107 is coupled through to the emitter of Q109. The pulse output of the Schmitt trigger (triggered by the sync signal) is thus coupled to Q109. This causes Q109 to turn on and Q110 to turn off and start the sweep just before it would normally begin. This is how the sweep is synced to the internal or external sync signal.

When the signal at the emitter of Q109 goes positive during retrace, a positive pulse is coupled via C107 to the base of Q108, the blanking amplifier. Because of the phase reversal of the common-emitter connected Q108, a **negative** pulse is coupled through C109 to the grid of the crt. This negative pulse cuts off the beam during retrace time, so that no retrace appears on the screen.

IO-102 Horizontal Amplifier

The horizontal amplifier of the Heath IO-102 is very similar to the vertical amplifier, as its schematic in Fig. 4-6

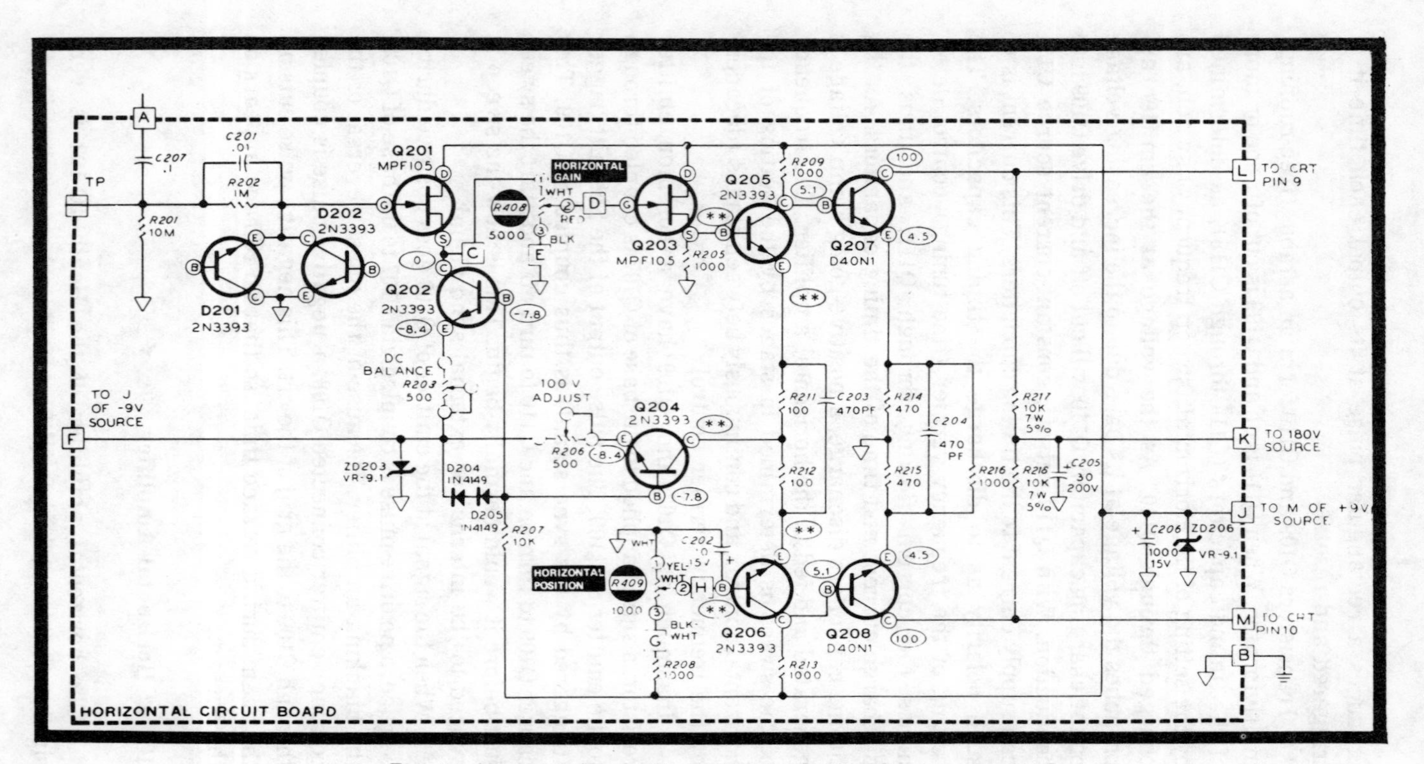

Fig. 4-6. Horizontal amplifier of Heath IO-102. (Courtesy Heath Co.)

shows. The main difference is that the horizontal amplifier does not have pnp amplifier stages corresponding to Q107 and Q108 in the vertical amplifier.

The requirements of the horizontal amplifier are not as stringent as those of the vertical amplifier. For one thing, not as much gain is required. The output of almost any sweep generator is on the order of a few volts, while the vertical input signal may be 30 mV or less. For another thing, not as great a high-frequency capability is required of the horizontal amplifier. Usually, at least two cycles are desired in a scope display, and this means that the frequency handled by the horizontal amplifier will be, at most, only half the frequency handled by the vertical amplifier. Also, the signal handled by the horizontal amplifier has a sawtooth waveform, which does not require as great a frequency range as may be required to amplify with fidelity some of the signals that might be applied to the vertical amplifier. The extra capabilities of the IO-102 horizontal amplifier beyond what is required should produce a very linear sweep.

The output of the horizontal amplifier is push-pull for minimum distortion. This is typical of the horizontal amplifiers of good quality scopes. If a single-ended horizontal output were used and the horizontal plates were nearest the electron gun, trapezoidal distortion in which the top and bottom edges of the display tended to converge might result.

In this horizontal amplifier, the positive-going ramp voltage of the sweep generator is amplified. It is then applied to the horizontal deflection plates of the crt. As the sawtooth voltage increases, it causes the electron beam to sweep across the screen, producing a visible trace. The sweep rate of the beam is determined by the sawtooth frequency.

IO-102 Power Supply

In Fig. 4-7 you see the schematic of the IO-102 power supply, and related circuitry including the voltage distribution network for the 5DEP1 crt. There are actually three distinct power supplies on the power supply circuit board. There is a high-voltage supply for the crt: D301, D302, and related

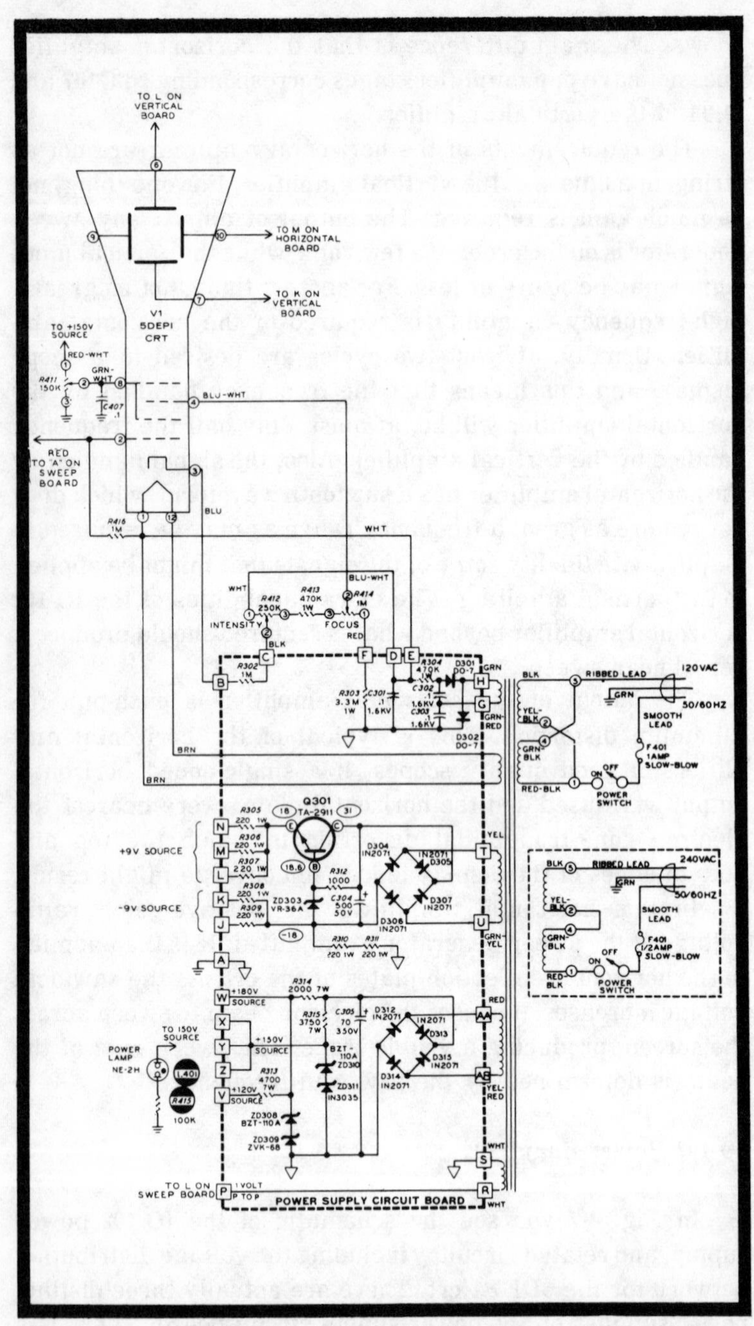

Fig. 4-7. The power supply and the voltage distribution network for the crt of the Heath IO-102. (Courtesy Heath Co.)

components. Below that in the schematic there is a regulated 9V supply for the low-level amplifiers. And at the bottom of the schematic, there is a power supply for the push-pull output stages of the vertical and horizontal amplifiers, and for the blanking amplifier.

Power line voltage is connected through the slow-blow fuse and on-off switch to the primary windings of the power transformer. The transformer has two windings, which may be connected in parallel for 120V ac operation in the USA or in series for 240V ac operation in some foreign countries.

The high-voltage winding of the power transformer is connected to the voltage doubler circuit consisting of D301 and D302, and C301 and 302. Capacitor C301 filters the negative high voltage (about −1500V), which is fed through R416 to the grid of the crt. The cathode is maintained about 50V less negative through the slider of intensity potentiometer (pot) R412. The second anode is maintained positive with respect to ground at a much lower voltage through the slider of astigmatism pot R411. R411 adjusts the roundness of the luminous dot on the screen. This control is adjusted so that the dc potential on the second (accelerating) anode is about the same as that on the deflection plates, about 100V. It is common practice to maintain the crt grid and cathode at a high negative potential and the anode much nearer ground. This of course results in the plate being highly positive with respect to the cathode, as required.

The first anode of the crt is maintained a few hundred volts positive with respect to the cathode through the slider of focus pot R414. Grid bias is adjustable between about 0 and 75 volts negative with respect to the cathode by intensity pot R412. The use of these and other controls will be fully explained later.

A low voltage secondary winding is connected to the full wave rectifier consisting of diodes D304 through D307 and C304. Zener diode ZD303 and resistor R312 maintain a constant voltage to the base of pass transistor Q301.

A simplified version of the low-voltage power supply is in Fig. 4-8. The unlabeled zener diodes in Fig. 4-8 are mounted on other circuit boards. A transistor connected in series with the

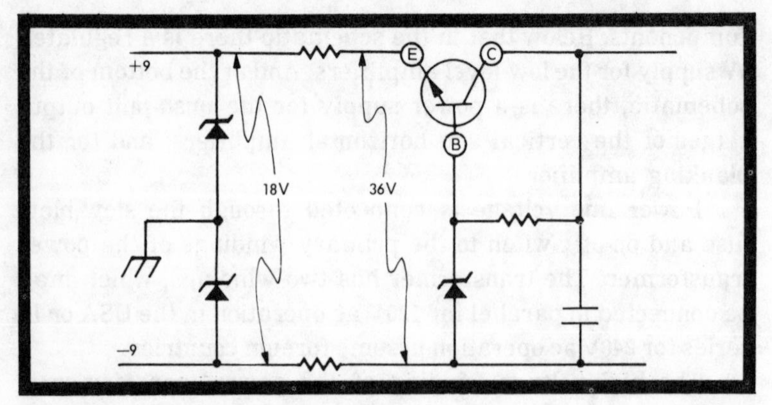

Fig. 4-8. Simplified schematic of the low voltage power supply of Heath IO-102 scope. (Courtesy Heath Co.)

power supply output, as Q301 is, is called a **pass transistor**, since it passes a controlled voltage to other circuits.

Refer again to Fig. 4-7. The output voltage of the low voltage power supply is regulated at 36V by series pass transistor Q301 and zener diode ZD303. Equal loads (not shown) are connected between each side of the supply and ground, so that two separate source voltages are obtained—one of +9V, and one of −9V.

Another secondary winding on the power transformer is connected to the full-wave rectifier consisting of diodes D312 through D315. Capacitor C305 filters the dc output. The full output voltage, 180V, is applied to the horizontal output amplifier. Zener diodes ZD308 and ZD309, and resistor R314, provide regulation of this voltage. Zener diodes ZD310 and ZD311, and resistor R315, provide a regulated 150V, which is applied to the astigmatism control. Resistor R313 and zener diode ZD101 on the sweep circuit board (Fig. 4-5) reduce the +180V to a regulated +120V for the blanking amplifier. Zener regulation of the power sources of the vertical and horizontal amplifiers and the sweep circuit results in an excellent display stability for the IO-102 scope.

Four printed circuit boards are used in this kit-type scope to minimize wiring and construction time. It is interesting to note that many laboratory scopes use a similar construction, and that the time-base and vertical amplifier circuits of these scopes are conveniently removable. When you buy one of these scopes, you buy separately the plug-in time-base and vertical amplifier suiting your particular requirements. Such modular construction is popular for the versatility and adaptability it gives a scope.

OSCILLOSCOPE PROBES

A probe is not, strictly speaking, a part of the scope circuitry. A probe is an extension of it, and is used for most measurements.

Ordinarily, the probe is connected to the vertical input by means of a coaxial (coax) cable. There are a number of reasons why an ordinary wire is not used for this purpose. For one thing, a stray capacitance would be produced between the wire and any metal object nearby, with the wire and metal object acting as capacitor plates and the insulation acting as the dielectric. The stray capacitance would vary as the wire was moved around, and the reactance would vary with it. High frequencies would be bypassed to varying degrees by the stray capacitances as the wire was moved, upsetting the measurements made. Another problem is so-called antenna effect: an unshielded scope lead would act as an antenna, picking up emi (electromagnetic interference) from any radio or TV working nearby. The displayed waveform would represent a combination of the desired signal and the interference. Of course, other types of emi might also be picked up by an unshielded lead.

Combination Attenuator and Frequency-Tuned Probe

A coax cable will eliminate emi being picked up by the lead, but the coax has a definite capacitance, so much per foot, which will tend to bypass high frequencies. A practical length of cable might bypass all frequencies above the audio range.

One probe circuit designed to minimize the effects of the capacitance of the coax is shown in Fig. 4-9. The coax picks up a voltage from the voltage divider R1-R2, which provides a 10-to-1 attenuation. This effect is taken into account in the attenuation ratio marked on the probe. If the probe is marked

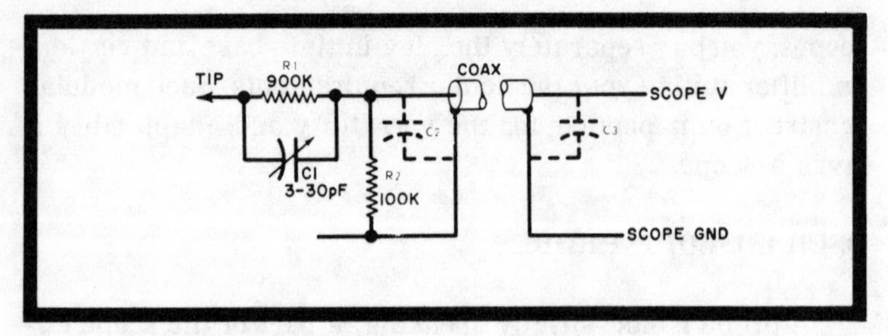

Fig. 4-9. Probe circuit designed to compensate for the capacitance of the coax lead and the scope input.

"10X ATTEN" you should multiply all scope voltage indications by 10.

The purpose of variable capacitor C1 is to minimize the effects of C2 and C3, which are, respectively, the capacitances of the coax and the scope input. These capacitances, C2 and C3, are in parallel with the input terminals of the scope, and tend to bypass high-frequency signals to ground. One problem that bypassing might cause, for example, is rounded corners on displayed square waves. Before the high-frequency components can be shunted by C2 and C3 in Fig. 4-9, they must pass through C1. So, as far as these signals are concerned, C1 is added in series with the undesired capacitances. Now, as you remember, capacitances in series add as resistances in parallel do, and since adding capacitances in series results in a **lower** total capacitance, C1 thus counteracts the undesirable effects of C2 and C3.

The value of C1 is made adjustable to compensate for varying lengths and other variations in coax cables, and for varying scope input capacitances. C1 is adjusted by the scope operator until the scope can display, with good fidelity, a square wave signal.

Demod Probe

A demodulator (demod) probe is used to examine modulation waveforms in the rf and i-f sections of radio and television receivers. Although the upper frequency limit of some scopes is only a few hundred kilohertz, these signals of perhaps hundreds of megahertz may be observed using a

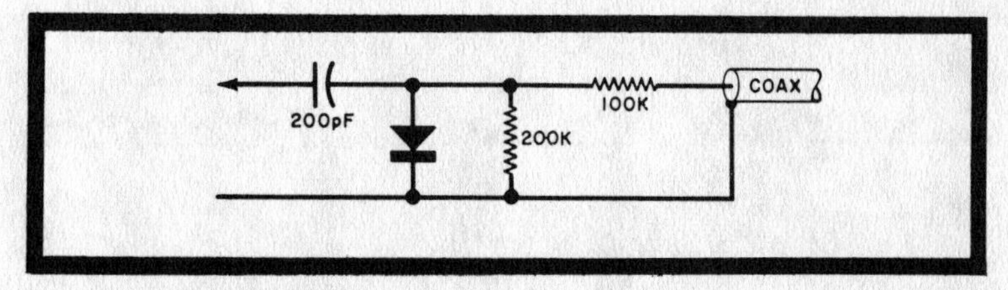

Fig. 4-10. Half-wave demod probe used for extracting the signal from the rf carrier for observation.

demodulator probe. The circuit illustrated in Fig. 4-10 is that of a half-wave type demod probe. Basically, the circuit is a broadband crystal detector with rf filtering on the output. Common commercial demod probes can easily extract the signal from a carrier up to 250 MHz.

Special Purpose Probes

There are probes for just about any conceivable application. The RCA WG-499A probe, for example, is specifically designed for 4.5 MHz trap alignment in TV sets. This probe, shown in Fig. 4-11, is an example of a **frequency-sensitive** probe. There are also high-voltage probes, some of which can measure the **high voltage** on a crt directly. Although the scope is basically a voltmeter, there are special **current probes** available that enable it to measure the alternating or direct current in a circuit without opening the circuit. Besides the **passive** probes described so far, there are **active** probes, which have an amplifier built right into them. These minimize the effects of emi picked up by the cable and minimize loading of a circuit under test.

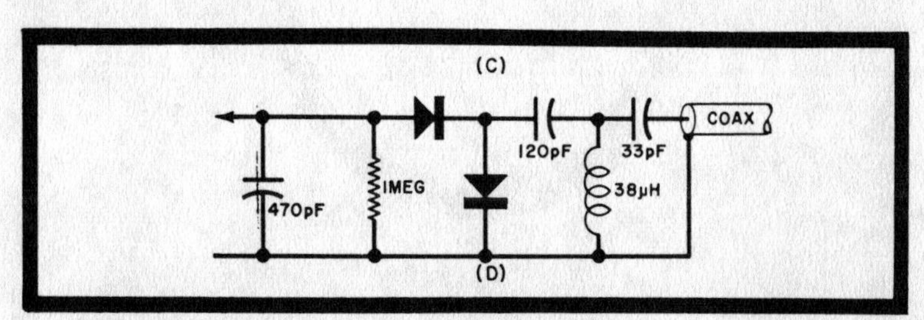

Fig. 4-11. RCA WG-499A probe is specially designed for use in 4.5 MHz trap alignment of TV receivers. (Courtesy RCA.)

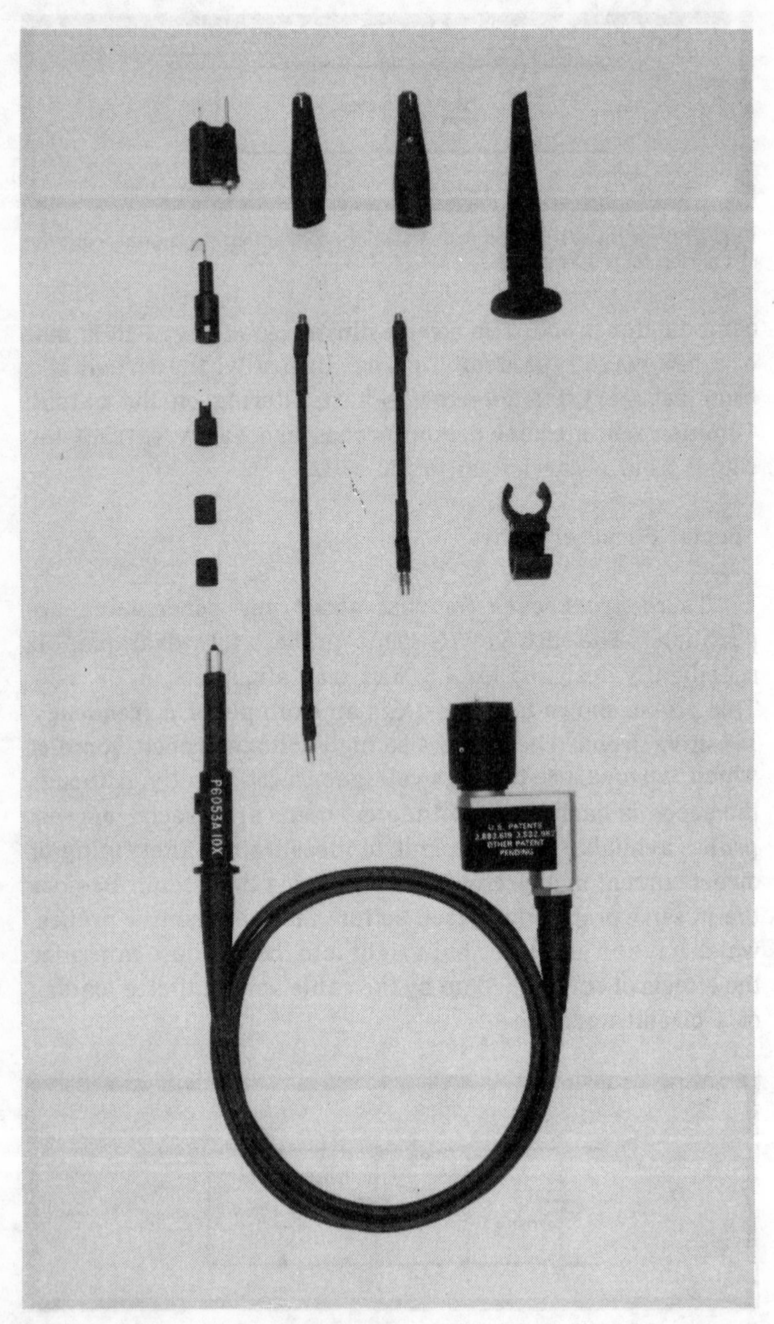

Fig. 4-12. Tektronix P6053A probe is an example of a x10 passive oscilloscope probe. (Courtesy Tektronix, Inc.)

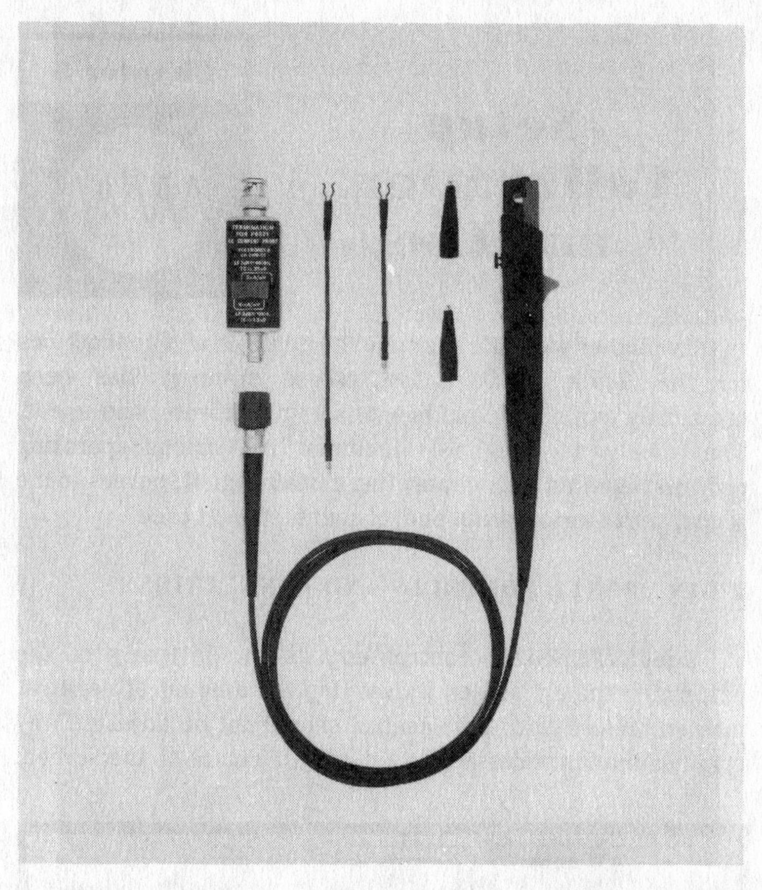

Fig. 4-13. Tektronix P6021 ac current probe and accessories. (Courtesy Tektronix, Inc.)

The Tektronix P6053A probe and some of its accessories are shown in Fig. 4-12. This is a miniature, high-frequency, 10 times **attenuation probe** intended for use with Tektronix 7000-series scopes. It may be compensated for use with all scopes having an input capacitance of 15 to 24 pF.

Another Tektronix probe, P6021 **ac current probe** shown in Fig. 4-13, obtains a signal in the matter of a clamp-on ammeter. To use the probe, one opens the spring-loaded slide, inserts the conductor, and releases the slide. A signal is electromagnetically coupled to a 125-turn pickup coil in the probe head, and therefore no direct electrical connection is required. Circuit loading, of course, is minimal.

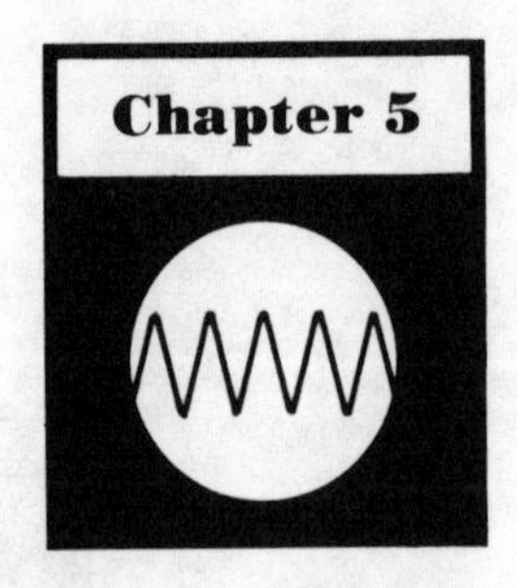

Setup Calibration, and Use

In this chapter we shall examine the controls and connections for the Heath IO-102 scope, whose circuitry has been previously explained, and how to set up, calibrate, and use it. The front panel (Fig. 5-1) includes the various operating controls, terminals for connecting certain signal sources to the scope, an indicator lamp, and, of course, the crt face.

FRONT PANEL CONTROLS AND CONNECTIONS

The INTENSITY control adjusts the intensity of the pattern on the crt screen by varying the amount of negative bias on the crt grid. This control should not be adjusted any brighter than is necessary to get a legible trace on the screen.

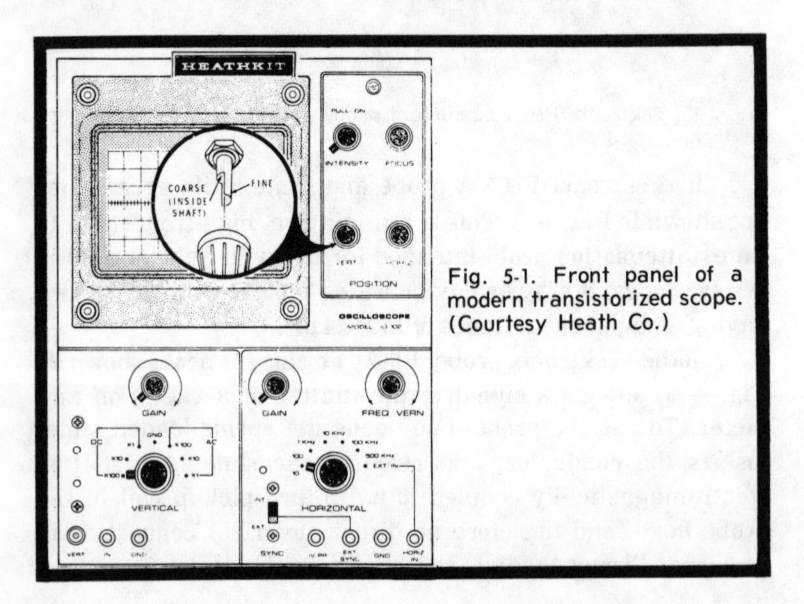

Fig. 5-1. Front panel of a modern transistorized scope. (Courtesy Heath Co.)

If the horizontal and vertical deflection voltages are removed, the intensity control must be turned all the way down so that the luminous spot where the electron beam strikes the screen is not visible. If a small, bright spot is concentrated for too long at any one point on the screen, the phosphor may be permanently damaged at that point. The switch that connects the scope circuitry to the power line, the **PULL ON** on-off switch, is associated with the intensity control. When ac power is applied to the scope, the power lamp comes on.

By adjusting the positive potential on the first anode, the **FOCUS** control determines the width of the electron beam, and thus the sharpness or clearness of the pattern on the crt.

The vertical **GAIN** control increases or decreases the vertical height of the trace by varying the gain of the vertical amplifier. This control is used in conjunction with the VERTICAL **attenuator.**

Signal attenuation is provided by the vertical attenuator switch. This switch determines what portion of the signal applied to the scope for observation is actually fed to the vertical amplifier. Each position of the attenuator represents a tenfold increase or decrease over the next one in the signal applied to the vertical amplifier. That is, changing the setting from x1 to x10 **decreases** the signal applied to the vertical amplifier by a factor of ten. Or, conversely, changing the setting from x10 to x1 **increases** the signal applied to the vertical amplifier by a factor of ten. Therefore, the vertical gain attenuator switch is analogous to the range switch of a voltmeter.

Another function of the vertical attenuator switch is to provide ac or dc coupling between a signal source and the scope. When dc coupling is desired, the vertical input coupling capacitor is bypassed. The attenuator switch also grounds the input circuit to provide a zero reference.

A **vertical dc balance control** on the bottom of the scope minimizes vertical movement (jitter) of the display when the vertical gain control is being adjusted.

The VERT POSITION control provides a controlled voltage imbalance in the vertical amplifier, enabling the entire pattern to be shifted upward or downward on the screen. Large adjustments of the VERT POSITION when first setting up the scope or when servicing it are performed by removing the knob, inserting a screwdriver into the hollow shaft of the control, and turning the screwdriver until the desired positioning is obtained. Fine adjustments of the vertical position of the trace during normal operation are made by turning the knob, which rotates the shaft itself.

Shifting the whole pattern left or right can be accomplished by rotating the HORIZ POSITION control. Acting similarly to the vertical position control in the vertical amplifier, the horizontal position control introduces a controlled voltage imbalance in the horizontal amplifier that results in the pattern being shifted horizontally. There is, however, no horizontal control corresponding to the coarse vertical position control.

The HORIZONTAL **frequency switch** determines the frequency range of the sweep generator. The total frequency range of the scope we are discussing is from 10 Hz to 500 kHz. This range is covered in five settings of the horizontal switch. With the horizontal switch set as shown in Fig. 5-1, the horizontal sweep is limited to frequencies **between** 10 Hz and 100 Hz.

The horizontal sweep frequency is continuously adjustable between the boundary frequencies determined by the horizontal frequency switch by varying the FREQ VERN (frequency vernier) control. This may be considered the fine frequency control, while the horizontal switch may be considered the coarse frequency control. The frequency vernier control is sometimes called the **sweep vernier**. With the horizontal frequency switch in the position shown in Fig. 5-1, the frequency vernier control can select any desired frequency between 10 Hz and 100 Hz. If the horizontal frequency switch were turned one position clockwise, any sweep frequency between 100 Hz and 1 kHz could be selected with the frequency vernier control.

Controlling the width of the crt display is the function of the HORIZONTAL GAIN control. This is accomplished by varying the gain of the horizontal amplifier, hence its name.

The INT-EXT switch selects either an internal or external

sync signal. When the switch is in the internal position, a sample of the signal being displayed is applied to the trigger circuit and used for synchronizing the sweep generator. When the switch is in the external position, a signal from some outside source is applied to the trigger circuit of the sweep generator, and the sweep is synced to this signal. The sync signal from an outside source is applied to the scope by means of the EXT SYNC connector.

The HORIZ IN connector, in the lower right corner of the front panel, allows the sweep signal to be applied by an external generator. The internal sweep generator is turned off when the horizontal sweep switch is in the **ext in** position, and the signal at the horizontal input connector is applied directly to the horizontal amplifier. The signal applied to the horizontal input connector may be a sine wave that is to be compared to a signal applied to the vertical amplifier. This is the way frequency is measured with a scope, as will be explained later.

The VERT IN connectors provide a means of applying to the scope the signal that is to be observed. The signal goes from the vertical input connectors to the vertical attenuator.

The **1V PP** connector is connected to a secondary winding of the scope's power transformer, and can be used to calibrate the vertical amplifier, or as a comparison to determine the value of a voltage of unknown amplitude. This connector simply provides a sample of the power line voltage being used to operate the scope. For calibration or for comparison usage, you simply use a coax from 1V PP to VERT IN, or practically speaking, insert the vertical input probe in the 1V PP jack.

TRYING OUT THE CONTROLS

In discussing the use of the scope, we are going to start right at the beginning. This will give you who have no hands-on experience with the scope an opportunity to start from scratch, while giving the person with some experience a chance to review the basic operating principles.

This chapter is written around the Heath IO-102 scope, but the instructions and information given should be readily transferrable to other scopes. You may not wish to actually perform the operations described while reading the text. You may still benefit from following the discussion, and where appropriate, referring to Fig. 5-1.

Initial Settings

The most certain way to learn the proper procedure for setting up the controls of a scope is to consult the manufacturer's instructions. In the absence of an instruction manual, the following procedure for setting the controls should be effective:

INTENSITY. Set fully counterclockwise (ccw).
VERT POSITION. Set near the middle of its range.
VERTICAL GAIN. Set fully ccw. The vertical gain is now turned all the way down.
VERTICAL **Attenuator**. Set for ac coupling in the x100 position.
HORIZONTAL **Frequency Switch**. Set to 10 to 100 position.
INT-EXT **Switch**. Set to INT.
HORIZONTAL GAIN. Turn fully ccw.
FREQ VERN. Set fully ccw.
HORIZ POSITION. Set to center of its range.
FOCUS. Turn fully ccw.

Off-On Switch. Turn on by pulling out the INTENSITY knob. The indicator light will also come on.

You may be wondering why the scope is turned on last instead of first. Since a solid-state scope such as the Heath IO-102 requires only a minute of warmup time, but there may be a slight tendency for 30 minutes or so for the display on the screen to drift up or down. You'd think it might be best to turn it on first so drift would be minimized. But do you recall, a small, bright, unmoving spot on the screen may cause permanent damage to the phosphor?

Now that initial settings have been concluded, we are ready to continue with the setup procedures.

Setup Procedures

1. Slowly turn the INTENSITY knob cw until a spot of light appears on the screen. Adjust the FOCUS control to make the spot as small and sharp as possible. Next turn up the HORIZ GAIN control until there is a horizontal line about 8 cm, i.e., 8 divisions long on the screen. Adjust the INTENSITY so the line is just bright enough to see plainly, and readjust FOCUS for the finest possible line. It is necessary to readjust FOCUS when you adjust INTENSITY; these two controls interact because they are both part of the same voltage divider circuit. The beam is focused in much the same way as a ray of light is focused by an optical lens system. Potentials on the crt electrodes focus the ray so that it comes to a fine point at the screen. These potentials are adjusted by the INTENSITY and FOCUS controls.

2. Connect a cable between the 1V PP terminal and VERT IN terminal. You are now applying a sample of the 60 Hz line voltage as an input signal to the scope. Now set the VERTICAL ATTENUATOR switch to whichever ac position results in a pattern that is 4 cm, i.e., 4 div p-p. Adjust FREQ VERN until a display of three cycles appears in a stationary position on the screen. The pattern on the screen, when made stationary, is said to be **locked in**. Notice that as you rotate the FREQ VERN control slightly to the left or to the right the pattern will start to drift horizontally. You now should have a feel for the operation of the vertical attenuator and frequency vernier controls.

3. Now vary the VERTICAL GAIN control and notice the effect of this on the display on the screen. Do the same with the HORIZONTAL GAIN control, and again note the effect.

The vertical gain control determines the amount of gain in the vertical amplifier. It determines how much the signal at the vertical input is amplified before it is applied to the vertical deflection plates. If the signal is not amplified very much, it will have a low height or amplitude on the screen. If the gain is turned way up, the signal applied to the deflection plates is much stronger and the display on the screen is much bigger vertically.

The horizontal gain control regulates the amount of amplification performed on the output of the sweep generator before it is applied to the horizontal deflection plates. This control varies the width of the display on the crt. It may be considered a width control, and the vertical gain control may be considered a height control.

Adjust the vertical and horizontal gains so that you get a presentation that is about 3 div high and 3 div wide. By varying the HORIZ POSITION control, move the pattern back and forth on the screen. Move the pattern up and down with the VERT POSITION control to get the feel of the control.

DISPLAYING THE OUTPUT OF AN AUDIO GENERATOR

An audio generator is an instrument used to generate ac signals for test purposes. Many such generators generate both sine wave and square wave signals, either separately or simultaneously. Sine waves generated by such instruments typically range from 20 Hz to 200 kHz. Square waves are usually generated over a somewhat narrower range. Waves supplied by the audio generator may be used for many testing purposes. We will discuss square wave testing later.

To make the image on the screen stand still, the horizontal sweep frequency must be set to a submultiple of the vertical signal frequency. As the sweep frequency is lowered, the electron beam takes more time to sweep across the screen. If the frequency of the sweep is reduced to half its former value, the beam will take twice as much time to sweep across the screen from edge to edge. If the sweep requires twice as much time as the input signal requires to complete one cycle, then two cycles of the input signal will be displayed on the screen. **The number of cycles in the display is equal to the signal frequency divided by the sweep frequency.** The higher the sweep frequency, the fewer cycles there are in the display. When the horizontal switch is set in the 10-100 position, the highest sweep frequency available is 100 Hz. If the input frequency is 5000 Hz, the **minimum** number of cycles of signal that can be displayed is 5000 divided by 100 equals 50. This is too many cycles for a legible pattern. Thus, to get a legible pattern, it is necessary to change the horizontal setting to the 100-1000 range. Let's see how this works.

We assume here that you've set the scope controls as described heretofore. If the 1V PP terminal is still connected to the vertical terminal of the scope, remove the connection. Now perform the following steps:

1. Set the audio generator up to supply a square wave signal of 5000 Hz, and connect the square wave output to the vertical input of the scope. Adjust the scope controls to obtain a waveform.

2. With the horizontal switch of the scope in the 10-100 position, set FREQ VERN to display as few cycles of the square wave signal as possible, in a stationary position on the screen. Notice that the minimum number of cycles obtainable is 50. To restate the formula: the input frequency divided by the highest sweep frequency in the range selected gives 50 cycles on the screen.

3. Move the horizontal switch to the 100-1 kHz position. You should now be able to obtain a stationary display of as few as 5 cycles.

4. Set the horizontal selector to the 1-10 kHz position, and you should be able to obtain one cycle in the display. One cycle will be displayed when the sweep horizontal sweep frequency is adjusted to match the input signal frequency. In this case, 5K divided by 5K equals unity. If the sweep frequency is increased still further, less than one cycle will be displayed, because the sweeping will occur at such a fast rate that less than one input cycle is completed during any given sweep cycle.

5. Now set the frequency vernier so that three complete cycles are in the display, and adjust the vertical gain so that the tops and bottoms of the cycles go off screen. Now rotate the vertical position control so that you can see the tops of the cycles again. Then rotate the vertical position in the other direction until the bottoms of the cycles are visible.

6. Adjust the vertical gain so that the p-p height of the display is 3 or 4 cm. Now turn up the horizontal gain so that the display is expanded and goes off the screen at both sides. Then adjust the horizontal position control so that the left side of the display is visible again.

Here is a special note for you to remember: although the vertical and horizontal amplifiers can handle large signals, too large a signal can cause **distortion**. Since the vertical amplifier is used to amplify the signal being observed, we must be careful not to overdrive it, or distortion seen in the display will have been created in the scope rather than coming from the equipment being tested.

CALIBRATING THE VERTICAL AMPLIFIER

A scope can be calibrated to permit its use for other ac or dc voltage measurements. To achieve this, some scopes employ a self-contained voltage calibrator, which supplies either a fixed or variable measured voltage to the input of the vertical amplifier. Other scopes, and this includes the Heath IO-102, usually provide a measured ac test signal voltage at a binding post on the front panel. When a scope has no internal means of calibration, some external means must be used. We'll see about dc calibration later.

AC Calibration

To calibrate a scope for ac, the controls are set for proper waveform viewing and the ac calibration voltage is then applied to the vertical amplifier. The magnitude of the calibration voltage determines the setting of the vertical attenuator. Suppose it is required to calibrate the scope for a vertical deflection of 1V per cm on the x1 vertical attenuator range, and suppose also that a 1V p-p calibration voltage is available, as on the Heath IO-102. Set the attenuator switch to the x1 range and adjust the vertical gain control for 1 cm of vertical deflection. **Make sure that the waveform is not distorted!** Distortion caused by overloading the vertical amplifier defeats the purpose of calibration. As long as the vertical gain control is not varied, the scope will now be calibrated so that 1 cm of vertical deflection on the x1 range corresponds to 1V p-p, 2 cm will correspond to 2V p-p, and so on. On the x10 range, 1 cm of deflection will correspond to 10V, etc. And on the x100 range, 1 cm will correspond to 100V. As you can now understand thoroughly, when the scope is used for voltage measurements, the attenuator is comparable to the range switch of a voltmeter. But unlike a voltmeter, the scope displays **amplitude** and **waveform**.

Now, to measure the p-p value of an unknown ac voltage, set the vertical attenuator to ac at highest range, apply the unknown voltage to the vertical input and then reset the vertical attenuator to a range that will result in a measurable, undistorted waveform. With the known calibration being 1V p-p equals 1 cm, you can easily see that the value of the signal will equal the p-p height of the waveform in cm times the setting of the attenuator. If the p-p height is 1.5 cm and the attenuator is set to x100, for example, then the value of the unknown voltage is 1.5 x 100 equals 150V. For practice, I suggest you imagine several other examples like this.

If an **external** calibrator with a variable measured ac voltage is available, a comparison method may be used for measuring an unknown ac voltage. In this method, the unknown signal voltage is applied to the vertical input of the scope and the vertical controls adjusted for a convenient deflection, say 2 cm. Without disturbing the scope controls, the output of the variable-voltage calibrator is substituted for the unknown voltage at the vertical input, and the calibrator controls are adjusted until the deflection due to the calibrating voltage matches the deflection that was observed with the unknown voltage. The output voltage of the calibrator, as read from its output control, equals the value of the unknown voltage.

If neither internal calibration nor an external calibrated variable-voltage source is available, the scope may be calibrated with a battery, or it may be calibrated with a step-down transformer such as a bell transformer.

DC Calibration

A good-quality dc scope may be accurately calibrated against a standard mercury cell. The terminal voltage of a mercury cell **voltage standard** stays very close to 1.357 volts over its lifetime. When a mercury cell is connected to the input terminals of a scope that is displaying an unvarying horizontal line, set for the sake of convenience at zero cm, the line will be deflected upward or downward, depending on the polarity with which the cell is connected. The amount of deflection corresponds to the voltage of the cell. Say it is indeed 1.357 cm. Now adjust the vertical gain control of the scope until the voltage of one mercury cell deflects the horizontal baseline **one division** on the scope scale, then two cells will cause a two-division deflection, three cells will cause a three-division deflection, and so on. And you could use the scope as a dc voltmeter, knowing that each vertical division equals 1.357V. This is of course a **dc** measurement. However, a **good** dc scope will give **the same deflection when an ac voltage having a p-p value equal to the dc reference voltage is applied.**

If you try this measurement, notice that no deflection is caused by the mercury cell if the vertical switch is in one of its ac positions, because a blocking capacitor is introduced into the vertical signal path when the vertical switch is in one of the ac positions.

The calibration of a scope might be referred to as its **vertical sensitivity setting**, since it is a measure of the ability of a signal applied to the vertical terminals to deflect the beam, and since it depends on the settings of the vertical amplifier controls. This is a little different from the vertical **input** sensitivity, which is one of its basic, inherent characteristics, and which is given in the specs provided by the manufacturer. The vertical input sensitivity is determined by the design of the scope; the calibration, or vertical sensitivity setting, is determined by the control settings.

Calculating Volts Per Division

One good **ac** voltage standard is an ordinary bell transformer having a secondary rms voltage of 6.3V. The p-p value of a sine wave voltage such as the secondary voltage of the bell transformer is 2.828 times the rms voltage. It turns out that 2.828 times 6.3V is close to 18V; so the secondary peak-to-peak voltage is 18 volts. If the secondary is connected to the vertical input of the scope, as in Fig. 5-2, and the vertical controls are

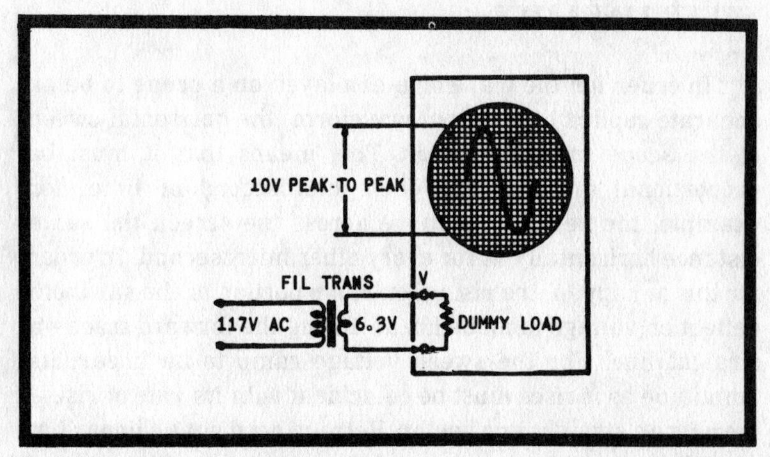

Fig. 5-2. An ordinary bell transformer provides a means of calibrating a scope for ac.

adjusted for a deflection of 3 cm, the calibration (the vertical sensitivity setting) is 18 volts per 3 cm, or 6 volts per cm.

The scope may be calibrated for any number of volts per division by setting the attenuator and vertical gain controls for the proper amount of vertical deflection as can be determined from the formula:

$$D_u = \frac{E_c}{\text{Cal in volts per div}}$$

where D_u is the required deflection and E_c is the voltage of the calibration standard.

Using the bell transformer as the calibration standard, let's use this formula and find the deflection required for a calibration of 9 volts per division:

$$D_u = \frac{E_c}{\text{Cal}} = 18/9 = 2\text{ div}$$

The divisions spoken of may be either centimeters or inches but most scopes have a centimeter graticule and cm is preferred.

SWEEP LINEARITY

In order for the waveform displayed on a scope to be an accurate replica of the input waveform, the horizontal sweep of the scope must be **linear**. This means that it must be proportional to time. For each microsecond of time, for example, the beam must move across the screen the same distance horizontally as for every other microsecond. In order for this to happen, the rising, or **ramp** portion of the sawtooth deflection voltage must be linear during the forward trace—a straight line. For the sweep voltage ramp to be linear, its amplitude as it rises must be coincident with its **rate** of rise—therefore a straight line vector. Retrace need not be linear but must be very abrupt (very high frequency).

Sweep linearity is especially important if the scope is to be used to make accurate time and phase measurements. We will describe a number of techniques for checking sweep linearity.

One thing that needs to be noted is that linearity may differ on each of the scope's sweep ranges. This means that a complete test of linearity requires a check of each range. Regardless of the range or sweep speed, the test method is the same.

Testing Sweep Linearity

Fig. 5-3 shows two cycles of a sine wave displayed on the face of a crt. Superimposed on this display is the etched grid, or graticule. Assuming all cycles of the test signal are identical, the cycles displayed should all have the same width. The pattern in Fig. 5-3 indicates that the sweep is linear.

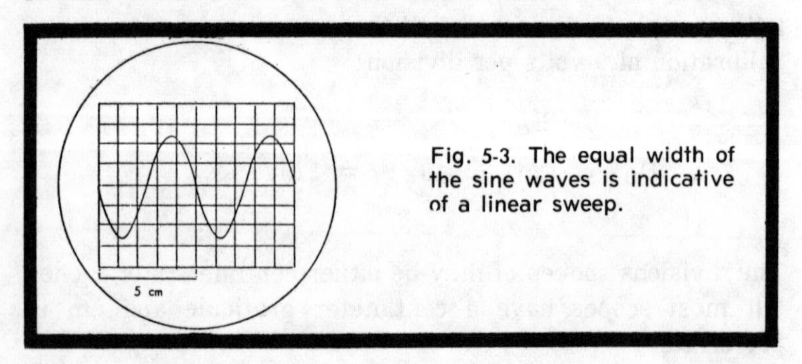

Fig. 5-3. The equal width of the sine waves is indicative of a linear sweep.

Spreading the width of each cycle by using the horizontal gain control makes possible a more accurate check. If the width of the first cycle is made equal to three horizontal divisions on the graticule, then every other cycle will have a width of three divisions if the sweep is linear. By adjusting the horizontal positioning control, the expanded pattern can be shifted left or right so that the width of each cycle can be measured and compared individually and consecutively against the original cycle.

Equivalent results can be achieved by applying a square wave to the vertical input of the scope and comparing the width of the cycles. This, of course, assumes that the square wave is symmetrical, that both alternations are the same width. If in doubt, expand one wave with horizontal gain, and use horizontal position to compare second waves.

Another technique involves converting the leading and trailing edges (the vertical sides) of a square wave into voltage spikes . The technique, called **differentiation**, amounts to applying a square wave to a series RC network, and taking the vertical signal from across the resistor. This signal consists of positive and negative pips corresponding to the leading and trailing edge of each alternation. If the spacing of the differentiated pulses is equal, the sweep of the scope is linear.

Yet another method of checking sweep linearity is possible if the scope has provisions for intensity modulation, or grid modulation. Since the axis of the crt that is parallel to the electron beam is designated the Z-axis, and since intensity modulation varies the strength of the beam itself, intensity modulation is also referred to as Z-axis modulation. To check sweep linearity, a differentiated square wave may be applied to the intensity modulation or Z-axis input. **No signal is applied to the vertical terminals in this method.** The trace on the crt is brightened by the differentiated pulses. By reducing the brightness of the trace using the brightness control the trace is then made to disappear so that only the intensified spots on the trace remain. If the horizontal distances between the bright spots are equal, the sweep is linear. Step-by-step procedures for checking sweep linearity by the various methods follow.

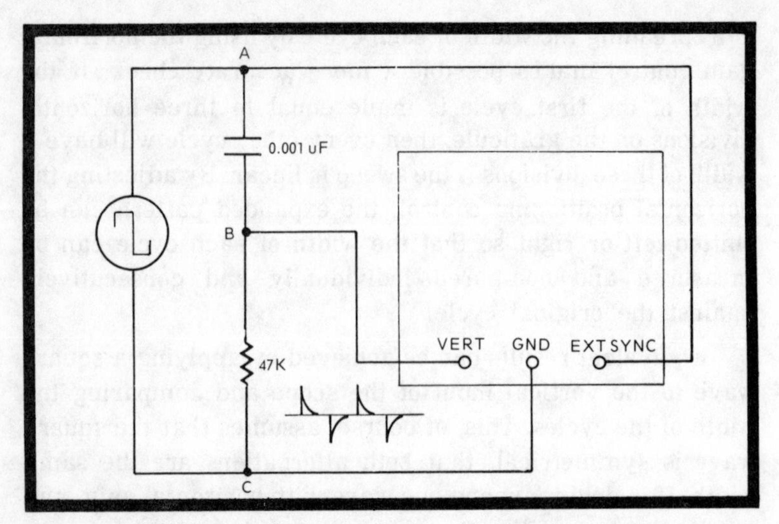

Fig. 5-4. In A, if the spacing between the pips of a differentiated square wave is uniform, the oscilloscope sweep is linear. B, test setup for sweep linearity test.

Audio Generator Method

1. Apply a 100 Hz sine wave from an audio generator to the vertical input of the scope and adjust the scope controls for a pattern 2 divisions high. Adjust the sweep controls so that three cycles of the sine wave are visible on the screen. The horizontal sweep control should be in its lowest range setting. Center the pattern so that the waveform is deflected one division above and one division below the X axis (the horizontal line that crosses the center of the screen). Now the display is 2 div p-p. Adjust the horizontal gain control until the display is three divisions wide.

2. Expand the pattern so that one cycle is two divisions wide. Recenter the pattern, if necessary, and measure and compare the waveform distances, by using the horizontal position control.

3. Reset the audio generator to 1 kHz and repeat steps 1 and 2.

4. Reset the generator to 10 kHz and repeat steps 1 and 2.

5. Reset the generator to 100 kHz and repeat steps 1 and 2. For most service-type scopes, you will have checked it on all of its sweep ranges, using the sine-wave method.

Differentiated Square Wave Method

1. Construct the circuit of Fig. 5-4 and apply a 10V 1 kHz square wave to the RC circuit, as shown. The ground lead of the square-wave generator is the one to be connected to point C (scope GND). Connect point A to the EXT SYNC terminal of the scope. Connect point B to the VERT IN (hot) terminal. Set the scope's sync switch to the external position. Adjust the scope controls until you obtain a display such as that in Fig. 5-4. The positive pips should extend one division above the X axis and the negative pips should extend one division below the axis. Expand the sweep with the horizontal gain control until the waveform distance is two divisions. Recenter the display, if necessary, and measure and record the distances of each cycle.

2. Check the sweep linearity at 100 Hz, 10 kHz, and 100 kHz. At the higher frequencies, it will be necessary to substitute smaller values for R and C of the series RC network to produce differentiated waveforms such as those in Fig. 5-4.

Intensity Modulation Method

1. If your scope has provisions for intensity modulation, reset the controls as in step 1 of the preceding method. Disconnect point B from the vertical input of the scope and connect point B to the Z-axis input. The crt trace will now be brighter at the leading edge of each pulse. Reduce the intensity of the trace until it disappears except for leading edges. Measure the distance between points and see if the distances are equal. If they are, the sweep is linear.

SYNCHRONIZATION (SYNC)

Unless the vertical input signal is synchronized with the horizontal sweep, the pattern will not stand still on the screen. If the frequency of the sweep generator is only two-thirds as great as the frequency of the displayed signal, each horizontal sweep of the beam will have a period equal to the time required for 1 and ½ signal cycles. Thus, during the first

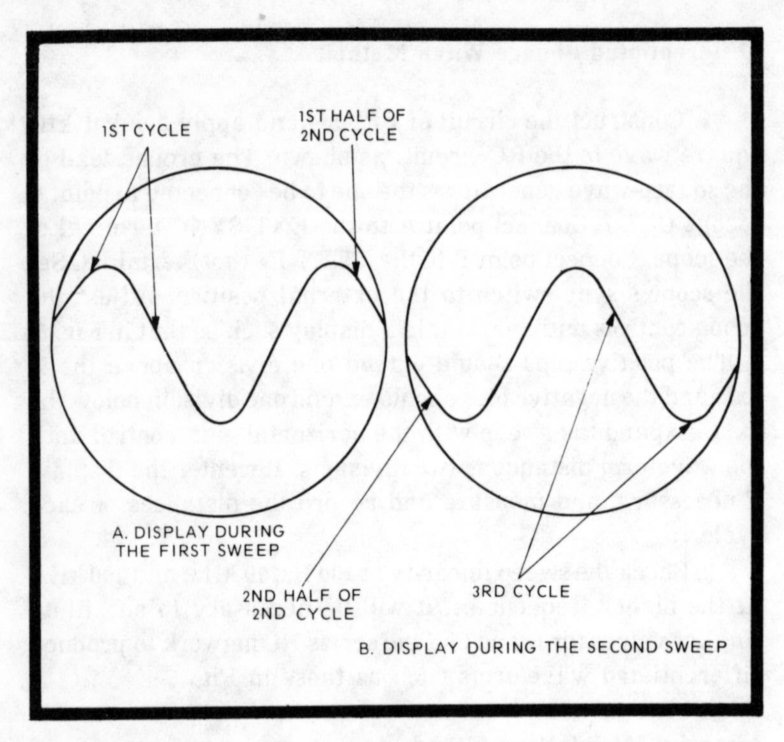

Fig. 5-5. When the sweep and signal are not synchronized, each sweep results in a different display, and the waveform drifts across the screen.

sweep from left to right, there will be 1 and ½ cycles of the signal displayed on the screen (Fig. 5-5A). At the beginning of the next sweep, the remaining half of the cycle that was only partially displayed during the previous sweep will be displayed, followed by another full cycle, as in Fig. 5-5B. Since the display is different for each sweep, the waveform will drift across the screen. All of this is due to the sweep and signal not being synchronized.

So that the display will remain stationary and properly observed, each successive trace of the electron beam across the screen must follow the same path, or **track**, as the one before. This means that the time required for one **sweep** cycle (sweep **and** retrace) must be the same as the time required for one or two or some other integral number of **signal** cycles. Under this condition, called **sync**, the electron beam begins each writing line at the same point on the signal waveform.

Sync Voltage Sources

In many scopes there are three different sources of sync voltage. The most commonly used source of sync is the displayed signal itself. The use of this source of sync is referred to as **internal sync**. When the INT-EXT sync switch is in the internal position, the signal itself is used as the source of sync.

Another source of sync is an **external sync generator** connected to the EXT sync terminal, with the INT-EXT sync switch set at EXT. The external sync generator may supply either a sine wave or a pulse signal for sync.

Another source of sync is the **power line voltage**, a sample of which may be available at terminals on the front panel of the oscilloscope, as the 1V PP in the Heath IO-102.

As we said, internal sync is the most commonly used. If, however, a sync signal of very accurate known frequency is required, then an external sync generator must be used. A very accurate known sync signal may be required, for example, for making frequency comparisons. Such a signal would be supplied by a very stable and accurately calibrated signal generator. The 60 Hz power line sync is used for special purposes.

Using Power Line Sample for Sync

The following procedure may be used to demonstrate the principle of synchronization generally, and use of a power line voltage sample particularly. If you don't have a scope, use **imagination**.

Apply a sample of the 60 Hz line voltage (available at the 1V PP terminal of the Heath IO-102 scope) to the vertical input of the scope. Set the INT-EXT switch to EXT. Set the horizontal sweep switch to the 10-100 position, and turn the FREQ VERN control until two or three cycles are displayed. Observe that it is **impossible** to obtain a stationary display. Now switch the INT-EXT switch to INT, and observe that a stationary display **can** be obtained. Turn the FREQ VERN control all the way cw, and then slowly turn it ccw. Try to obtain a display of six cycles, then five, four, three, two, and finally, one cycle.

Discussion. The free-running frequency of the sweep generator can be adjusted to any desired frequency using the horizontal frequency switch (coarse) and FREQ VERN controls. However, both the sweep frequency and the test signal will drift. Since they will not drifty by the same amount, owing to time delays within the scope itself, the displayed presentation will drift one way or the other as long as the sweep generator is operating in the free-running mode.

To synchronize the frequency of the sweep generator with the signal frequency, a portion of the test signal is used to trigger the horizontal generator. This causes the sweep generator to operate at a slightly higher frequency called the **triggered frequency**, or the **synchronized frequency**. Thus, the sweep frequency is able to keep pace with the signal frequency when a sync signal is applied to the horizontal generator. As we have noted, when the INT-EXT switch is in the EXT position, no sync signal is applied to the horizontal sweep generator, and the sweep and signal frequencies get out of step, making it impossible to obtain a stationary display. But when the switch is in the INT position, part of the signal is used to synchronize the sweep generator, to cause it to operate at a triggered frequency. When the triggered frequency is set to the same frequency or to a submultiple of the signal frequency, the display is stationary, or locked in. Turning the FREQ VERN control to the left (in the IO-102) increases the triggered frequency, so that fewer signal cycles are displayed on the screen during one sweep. For this reason, you were able to lock in the display at six, five, four, three, two, and one cycle by starting at the full cw position of the FREQ VERN and slowly rotating it ccw.

Disconnect the 1V PP terminal from the vertical input and connect it to the EXT sync terminal. Set the INT-EXT switch to the EXT position so that the 1V PP 60 Hz power line frequency is applied to the horizontal sweep generator as a sync signal. Apply a 1 kHz square wave signal from an audio generator to the vertical input terminals of the scope, and set the horizontal switch to the 100 Hz-1 kHz position. Now try to lock in the presentation. You will find that you cannot lock in the display with the INT-EXT switch in the EXT position, even though a 60-Hz sync signal is applied to the EXT SYNC terminal. Set the switch to the INT position, and lock in the pattern. Vary the setting of the FREQ VERN control and notice that it is possible to thus lock in different numbers of cycles on the screen.

Discussion. Since the input signal in this step was not 60 Hz nor a multiple of 60 Hz, the 60 Hz sync signal was unable to synchronize the sweep. When the INT-EXT switch was changed to the INT position, a portion of the 1 kHz input signal itself was substituted for the 60 Hz signal, and it was able to synchronize the sweep, giving a stable pattern. (THIS POINTS UP THE FACT THAT IN ORDER FOR SYNCHRONIZATION TO OCCUR, THE SIGNAL APPLIED AS SYNC MUST BE A SUBMULTIPLE OF THE VERTICAL SIGNAL, OR MUST EQUAL IT.

Set the INT-EXT switch to the EXT position again, and apply a 600 Hz square wave to the vertical input of the scope. Now try to lock in the display. Vary the frequency of the sweep by rotating the FREQ VERN control, and in this manner change the number of cycles in the display.

Discussion. You're right! With a 600 Hz input signal applied to the vert input, the 60 Hz EXT SYNC signal was able to synchronize the sweep. It was able to synchronize the sweep at a number of different sweep frequencies that are multiples of 60 Hz. THIS DEMONSTRATES THE FACT THAT AN EXTERNAL SYNC SIGNAL THAT IS HARMONICALLY RELATED TO THE VERTICAL SIGNAL CAN BE USED TO SYNCHRONIZE THE SWEEP FREQUENCY AND LOCK IN THE DISPLAY.

LISSAJOUS FIGURES

One way to determine the frequency of an **unknown** sine wave is to apply the unknown signal to the **vertical** input and a **known** sine wave frequency to the **horizontal** input. The resulting figure on the screen is called a **Lissajous** (Liss-a-Jew) figure, and from it one can obtain the frequency of the unknown sine wave.

The simplest Lissajous figure is a circle. See Fig. 5-6. This pattern is produced by two signals having the same frequency and amplitude, but differing in phase by 90 degrees. Fig. 5-6 shows how the two voltages applied to the vertical and horizontal deflection plates of the crt generate a circular pattern. Corresponding points in time bear the same number on each waveform in Fig. 5-6. The small numbers on the waveforms could represent microseconds, for example. Then at 4 usec after the starting time, the voltage on the vertical plates would be at its maximum, and at 7 usec after the starting time the voltage on the horizontal plates would be at its maximum. At any given time, the combined effects of the vertical and horizontal deflection voltages put the electron beam at a certain spot on the circle. At 4 usec, the combined effect is such as to put the beam at the top of the circle, for instance.

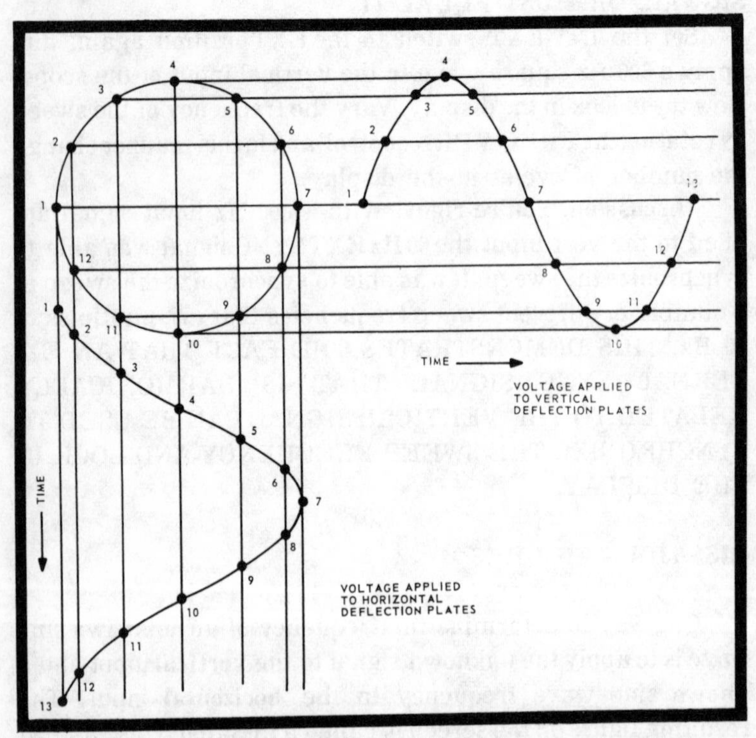

Fig. 5-6. A circular Lissajous pattern is produced by two signals having the same frequency and amplitude, but differing in phase by 90 degrees. (Reproduced from an Air Force technical manual.)

Lissajous patterns obtained when two different frequencies are applied to the vertical and horizontal deflection plates are illustrated in Fig. 5-7. These patterns result when the vertical frequency is **twice** the horizontal frequency. It doesn't make any difference just what the vertical and horizontal frequencies are. If their **ratio** is 2:1, the

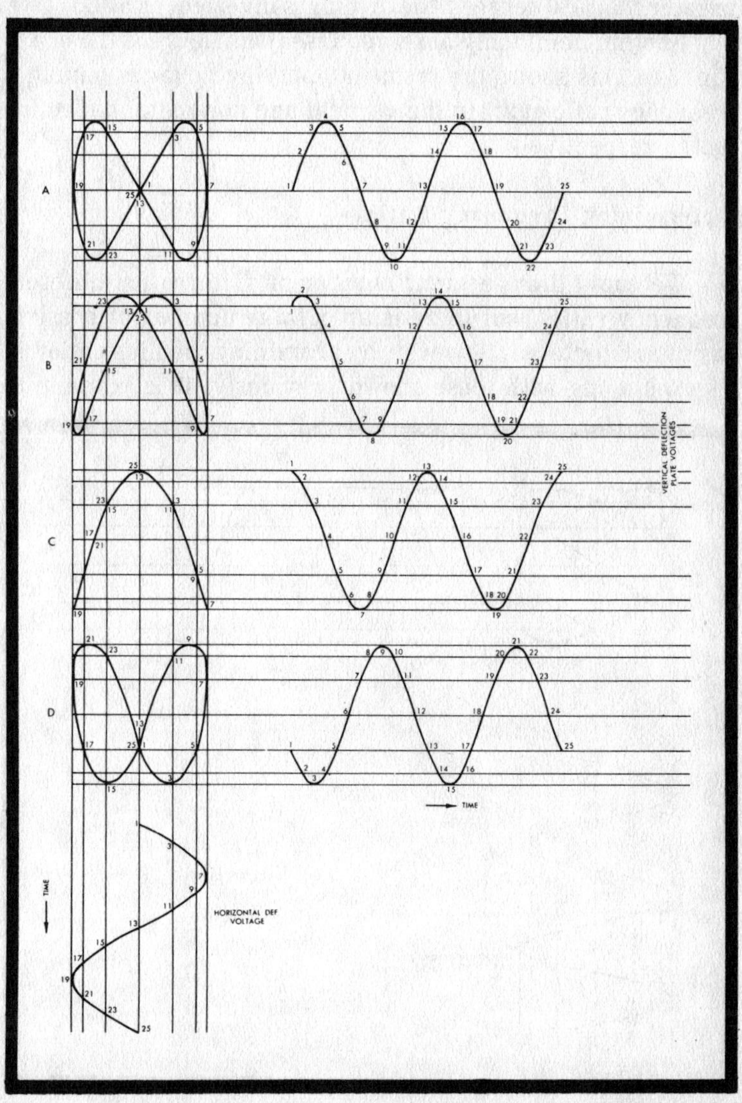

Fig. 5-7. Bowtie-shaped Lissajous patterns result when there is a 2:1 relationship between the vertical and horizontal deflection frequencies.

bowtie figures of Fig. 5-7 will result. Notice how the phase difference between the vertical and horizontal signals changes the Lissajous figures obtained with a 2:1 frequency ratio.

Two voltages in phase result in the neat bowtie shown in Fig. 5-7A. When the phase angle is 90 degrees, the loops are closed, as in C. This figure is a **parabola**. For phase angles greater than 180 degrees, the display is inverted, as at D.

Another commonly observed Lissajous figure is the one in Fig. 5-8. This shows the result of applying voltages having a frequency ratio of 2:3 to the vertical and horizontal deflection plates respectively.

Determining Frequency Ratios

We could draw a great number of pictures for different frequency ratios, but there is an **infinite** number of possible ratios and pictures. However, by examining the illustration in Fig. 5-8 along with those shown previously, it is possible to

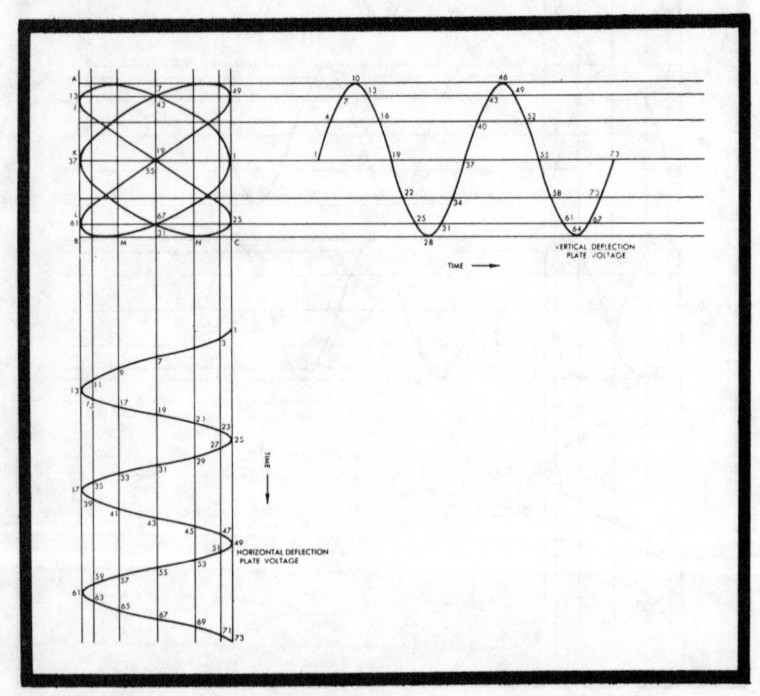

Fig. 5-8. The Lissajous figures obtained with a 2:3 frequency ratio. (From an Air Force technical manual.)

deduce a general rule by which we can determine any frequency ratio from the pattern it generates. In this regard, notice in the illustration of the 2:3 ratio figure (Fig. 5-8) that the pattern touches the horizontal line BC at **two** points M and N, and the vertical line at **three** points, J, K, and L. From these observations, you can form the following general rule: THE RATIO OF THE NUMBER OF POINTS AT WHICH A LISSAJOUS FIGURE TOUCHES A HORIZONTAL LINE TO THE NUMBER OF POINTS AT WHICH IT TOUCHES A VERTICAL LINE IS EQUAL TO THE RATIO OF THE VERTICAL FREQUENCY TO THE HORIZONTAL FREQUENCY. And the horizontal line can be top **or** bottom, the vertical, left **or** right.

In applying the rule to the pattern at C in Fig. 5-7, note that spot passes through point 25 (or, same thing, point 1) twice. Thus, although the pattern appears to touch the horizontal line only once, it actually touches it twice. Another thing to remember in connection with the rule is that it holds for any combination of vertical and horizontal lines as long as they intersect the pattern.

Determining the Frequencies

For accurate determination of frequencies by the Lissajous-figure method, you need a calibrated frequency generator. You apply the output of this generator to one set of deflection plates and the unknown frequency to the other set. Usually, the unknown frequency is applied to the vertical plates. Then you determine the frequency **ratio** from the rule just given. Read the frequency of the calibrated generator and use the ratio to determine the unknown frequency. If the vertical to horizontal ratio is, for example, 2:1 and the horizontal frequency supplied by the calibrated generator is 1000 Hz, the vertical, or unknown frequency is two times 1000 Hz. It is, in other words, 2000 Hz.

Here's how to observe with a scope the principles of Lissajous figures. Refer to Heath IO-102, Fig. 5-1.

1. Turn on the scope and the sine-square wave audio generator and let them warm up. Connect the sine wave output

of the audio generator to the vertical input terminals of the scope. Connect the 1V PP 60 Hz test signal from the front panel of the scope to HORIZ IN and set the horizontal range switch to EXT IN so that the 60 Hz signal is applied to the horizontal amplifier.

2. Adjust the horizontal and vertical gains for a display about 4 cm high and 4 cm wide. Adjust the frequency of the audio generator through the range 100 Hz to 140 Hz. Notice the pattern you obtain when the generator is set to 120 Hz. Adjust the sine-square wave generator to give a stationary display.

3. Set the frequency of the audio generator to 90 Hz. The ratio of the frequency on the vertical plates to that on the horizontal plates is 3:2. Thus, the Lissajous figure should touch a line at its side at two places and a line at its top or bottom at three places.

4. Set the frequency of the audio generator to 20 Hz and 40 Hz, in turn, and note the pattern you obtain each time.

5. Set the audio generator to 200 Hz, then slowly increase the frequency. Try to get ratios of 5:1, 7:1, and 9:1. These will be obtained when the settings of the generator are 300 Hz, 420 Hz, and 540 Hz, respectively.

Discussion. When the audio generator frequency was less than 60 Hz, the loops appeared on the sides of the pattern. When the generator was set to frequencies higher than 60 Hz, the loops appeared at the top and bottom. The orientation of the pattern depends on which is greater, the vertical frequency or the horizontal frequency. No matter what the orientation, the frequency ratio can still be determined from the Lissajous figure, using the general rule given earlier. However, the unknown frequency is generally applied to the vertical input. This is partly a matter of convention. Also, since the maximum frequency of the vertical amplifier is sometimes greater than the maximum frequency of the horizontal amplifier, this connection will sometimes permit the measurement of higher frequencies than could otherwise be measured.

Determining Phase Relationships

Lissajous figures are also useful for determining phase relationships between two voltages **of the same frequency**. This means that they are useful for determining the phase shift that might occur, for example, in an amplifier. In evaluating the performance of an amplifier, one is not concerned with the absolute phase shift of the signal in going through one or more stages of amplification, but rather with the difference in the phase shift at different frequencies. This is mostly a check on the low-frequency performance of the amplifier, since it is at the low frequencies that phase shift is likely to be greatest.

Theoretically, an amplifier stage either shifts a signal 180 degrees, or not at all. A complete amplifier, then, shifts an input signal by an exact number of multiples of 180 degrees, in theory. In practice, however, the amplifier may actually cause a net shift of a few degrees more or less, depending on its design and condition. If **all** frequencies were shifted by the same amount, the shift would not be noticeable, and it might or might not be harmful, depending on the application. But since the phase shift is frequency-variable, the result is distortion. Phase shift curves are often plotted against frequency. The significant thing about such graphs is not the amount of phase shift, but how it varies with frequency. A flat curve of phase shift vs frequency indicates good amplifier performance.

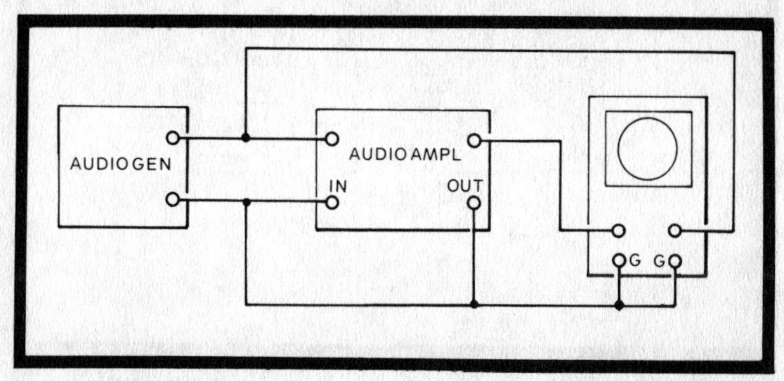

Fig. 5-9. Test setup for measuring the phase shift of an amplifier by Lissajous technique.

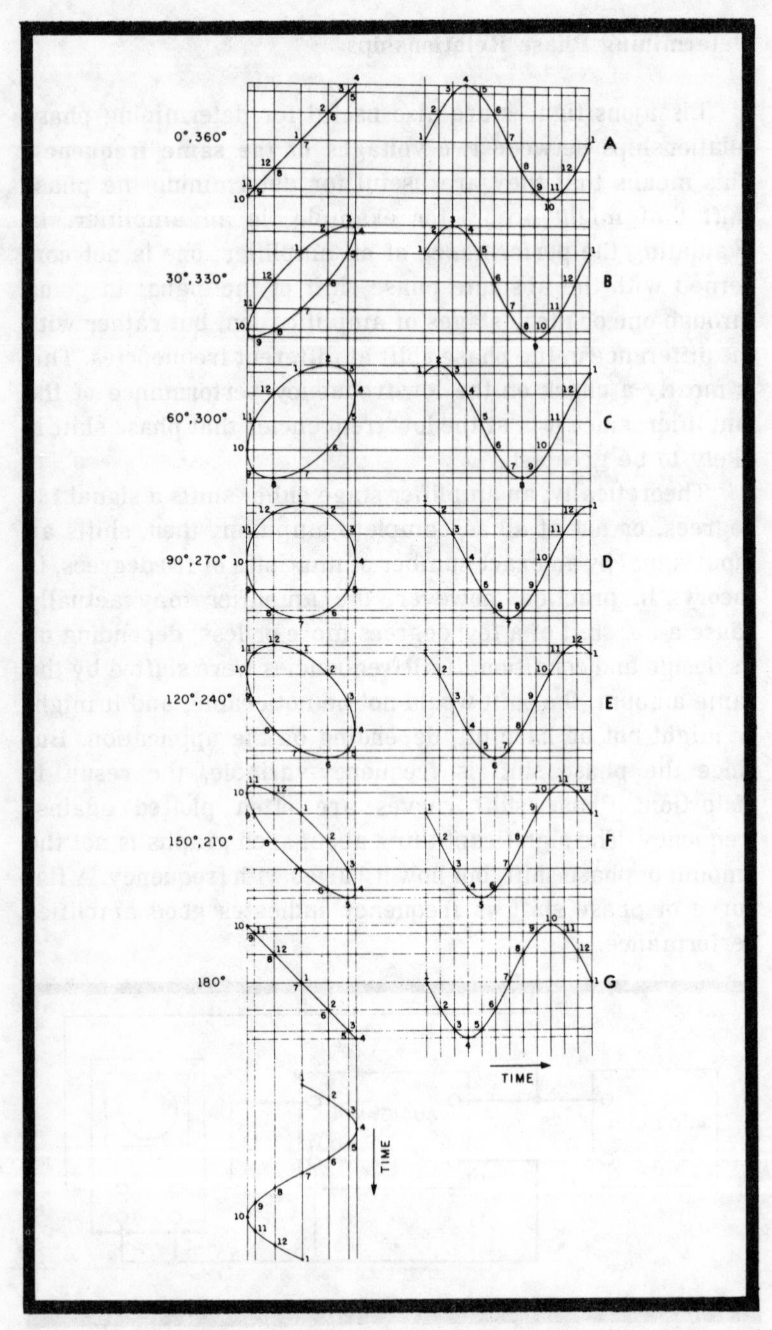

Fig. 5-10. Illustrating the effect of phase variation on pattern shape and orientation. (From an Air Force technical manual.)

Fig. 5-9 illustrates the Lissajous techniques for measuring the phase shift of an amplifier. The audio generator output is fed to the input of the amplifier at a level below that which would produce distortion due to overloading. The same input is also connected to the horizontal input of the scope with the internal sweep disconnected. (In the Heath IO-102, this is accomplished when the horizontal switch is in the ext in position.) The output of the amplifier is connected to the vertical scope input. The horizontal- and vertical-input frequencies have a 1:1 ratio; that is, they are the same. Vertical and horizontal gain are adjusted for a convenient-sized display. Fig. 5-10 gives a number of examples of phase-shift patterns obtainable. The 0- or 180-deg pattern in A indicates an ideal, which is seldom realized in practice. A 15-degree shift is more often attainable, and it is hardly noticeable. You can determine the approximate phase shift by comparing the results you obtain with those shown in Fig. 5-10. Figs. 5-10B through G have not been described—this deliberately so that you can study, think, and test your understanding.

A more accurate way of evaluating the Lissajous figures you obtain for phase shift is shown in Fig. 5-11. In using this method, first make certain that the figure is centered on the screen. Then measure **m**, the distance to the intercept on the vertical axis; and **n** the maximum vertical deflection from the horizontal axis. The phase angle is the angle whose sine is **m** divided by **n**. That is, θ°= arc sin(**m—n**) where θ is the phase

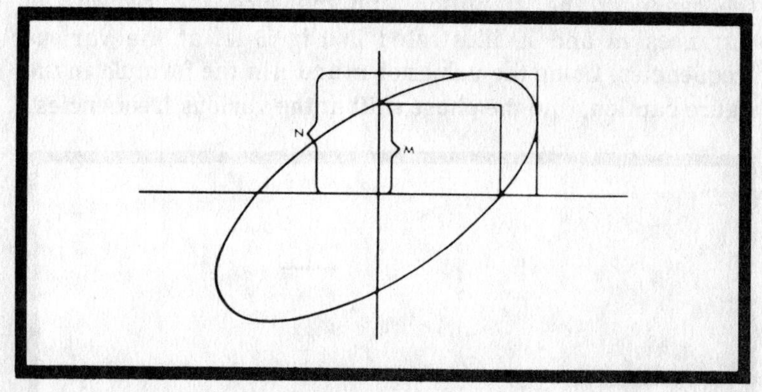

Fig. 5-11. The phase shift indicated by an elliptical Lissajous figure is given by the formula θ arc sin (m—n).

angle. If, to give an example, **m** over **n** equals 0.5, the angle must be 30 deg, because 30 deg is the angle whose sine is 0.5, as determined from a sine table.

Procedure for Measuring Amplifier Phase Shift

The discussion previously as to measuring amplifier phase shift was very general. The procedure following is exact and practical.

1. Turn on the amplifier, scope, and audio generator and allow them to warm up. **Turn off the scope sweep.**
2. Position the luminous spot at the center of the screen.
3. Connect the audio generator to the horizontal input of the scope and to the input of the amplifier, as shown in Fig. 5-9. Connect the output of the amplifier to the vertical input of the scope.
4. Adjust the vertical gain for a convenient height, as in Fig. 5-12. Note the number of graticule divisions occupied by the trace, then reduce vertical gain to zero, so that you again have a dot at the center of the screen.
5. Adjust the horizontal gain for the same length of line you obtained in the preceding step. See Fig. 5-12B. Now restore the vertical gain to the value used in the preceding step. If the phase shift of the amplifier were 90 or 270 degrees, a perfectly round Lissajous figure would now be seen on the screen.
6. Tune the audio generator to various frequencies over the range of the amplifier, and measure and record the distances **m** and **n**, illustrated in Fig. 5-11, at the various frequencies. Using the values of **m** and **n** in the formula in the figure caption, find the phase shift at the various frequencies.

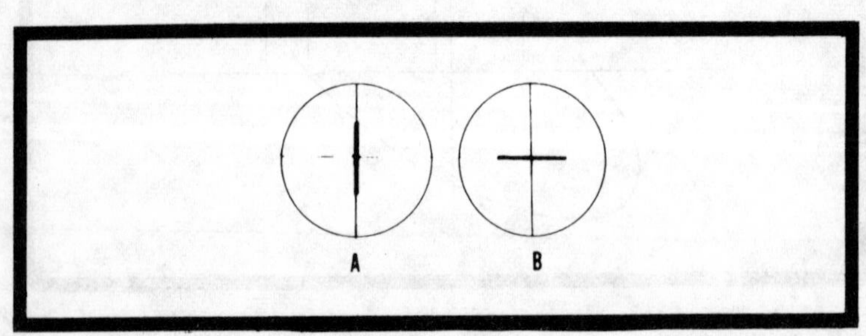

Fig. 5-12. Centering the cathode ray beam for phase measurements.

7. Plot a graph of phase shift against frequency. Mark the vertical scale in degrees of shift and the horizontal scale in kHz, or frequency. If the graph that results is almost a straight line, you've got a fantastic amplifier, as far as phase shift is concerned.

DIRECT PHASE MEASUREMENT

Phase angles between waveforms having the same frequency but appearing at different points in a circuit may also be measured directly, using external sync. For example, the phase angle between the voltage across a resistor and the voltage across a capacitor in a series RC circuit (Fig. 5-13) can be found by synchronizing the scope externally with the sine wave applied to the RC circuit. Through the use of external sync, the scope sweep is triggered at the same time no matter where the waveform is being observed. It may be observed across the capacitor, as in Fig. 5-13A, or it may be observed across the resistor, as in Fig. 5-13B.

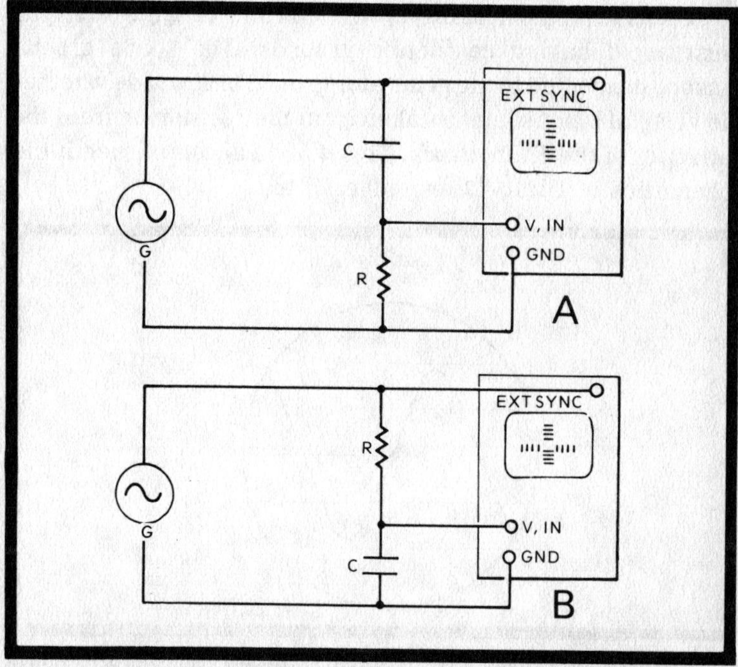

Fig. 5-13. A method of direct phase measurement.

In using this method, the audio generator is first connected directly to the vertical input of the scope to produce a reference for phase measurements. The reference sine wave display is then adjusted to a convenient width, perhaps 4 cm, and centered on the graticule so that the starting point of the sine wave lies on a vertical line A, as in Fig. 5-14.

Provided the scope sweep is linear, equal distances on the horizontal axis correspond to equal angular displacements. In Fig. 5-14, the 2 cm occupied by an alternation (point A to B) represents 10 small divisions. Since an alternation also represents 180 deg, it follows that each small division on the horizontal axis represents an angular displacement of 18 deg.

After the horizontal axis has been thus calibrated by means of the reference waveform, the vertical input leads of the scope are connected in turn across the capacitor and the resistor of the RC circuit, as in Fig. 5-13A, then B. The angular displacement of the waveform across the capacitor or resistor from the reference waveform is measured on the calibrated horizontal axis.

CAUTION: In hooking up the circuits of Fig. 5-13, you must avoid having conflicting grounds. The scope ground must be connected to the generator ground or low side whether the vertical input signal is taken from the resistor or from the capacitor of the RC network. This will be accomplished if the schematics of Fig. 5-13 are adhered to.

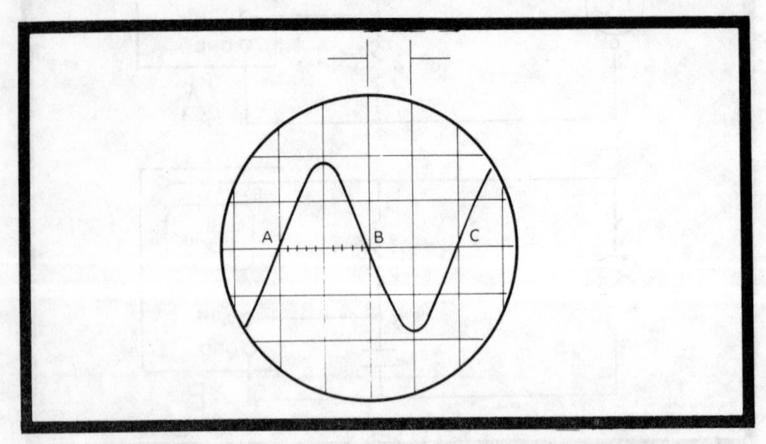

Fig. 5-14. Calibrating the horizontal sweep for direct phase measurement.

Here is the step-by-step procedure for the test just outlined.

1. Connect the hot lead (the high side) of the generator to the vertical input (hot) jack of the scope, and the ground lead (low side) of the generator to the scope ground. Short the EXT SYNC jack of the scope to the vertical input (hot) jack, and set the sync selector to the EXT position.

2. Apply a 2V p-p sine wave of 1000 Hz to the scope. Adjust the scope controls for three sine-wave cycles exactly centered on the X and Y axes, as in Fig. 5-15. Adjust the horizontal gain until the width of the cycle ABC is 4 cm, as in Fig. 5-14. This is the reference waveform, and the horizontal axis of the graticule is now calibrated in degrees and in terms of this waveform.

3. Without changing any generator or scope controls, remove the short from the vertical input to the EXT SYNC jack, and make the connections shown in Fig. 5-13A. For R, use a -33K resistor, and for C, use a 0.01 uF capacitor. Measure the displacement of the resulting waveform from point A, the starting point of the reference waveform. Remember that each minor division or calibration mark on the horizontal axis represents 18 deg. As you may know, measuring across C, the phase angle will be **lagging** Notice whether the start of the new wave is to the right or to the left of point A in Fig. 5-14.

4. Now connect the circuit as shown in Fig. 5-13B. Again measure the phase angle and notice whether the waveform is to the left or to the right of the reference waveform. This time

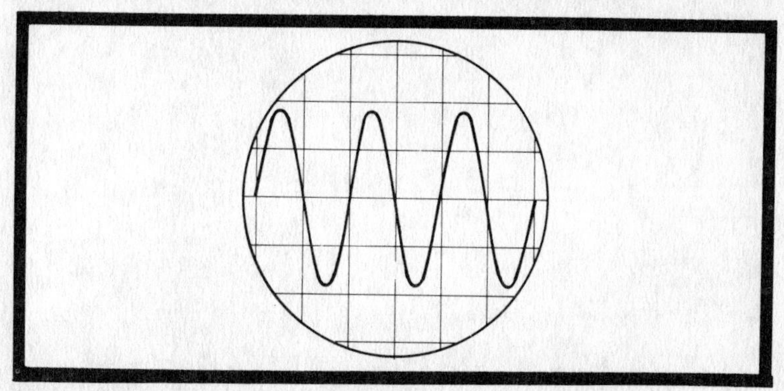

Fig. 5-15. Three sine wave cycles exactly centered on the X and Y axes in preparation for phase measurement .

the phase angle is **leading**, since the voltage is being measured across the resistor.

5. Determine the reactance of the capacitor at 1000 Hz, the frequency applied to the capacitor, according to the formula:

$$XC = \frac{1}{2\pi fc}$$

Compute the phase angle according to the formula:

$$\theta = \text{arc tan}\left(\frac{XC}{R}\right)$$

Or, draw an impedance triangle and solve for the angle graphically. Let R be the horizontal side of a right triangle and XC the vertical side, and draw the triangle to scale, so that the lengths of R and XC are in the proper proportion. The phase angle is the angle between R and the hypotenuse, and may be measured with a protractor.

6. Repeat the above procedure with the audio generator set to deliver an output of 500 Hz. Compare results.

Scope Versatility in Testing

In the preceding chapter we discussed the basic principles of operating a scope. We showed how the theory of operation of a scope is related to its use. In this chapter, we shall demonstrate its usefulness as an electronic measuring instrument, and shall discuss some of the more practical and sophisticated capabilities of the instrument.

SQUARE WAVE TESTING

In using a scope to test an amplifier or other circuit, the operation of the circuit is checked by comparing the output waveform to the input waveform. In this method, a pure sine wave or square wave is supplied to the input, and the output waveform is compared to this pure form. It is very important to remember that the waveforms being observed may be adversely affected by improper amplification in the vertical amplifier of the scope itself.

Any distortion caused by the vertical amplifier will appear on the screen just as if it were present in the signal applied to the scope. Distortion produced in the vertical amplifier can be quite misleading. For example, a scope might have a vertical amplifier with a frequency response of 100 Hz to 50 kHz. If this scope is used to test a circuit for its response at 50 Hz, a test signal which remained clean all the way through the rest of the test circuit would be distorted in the vertical amplifier. The distortion observed on the screen would appear to have been caused by the equipment under test, where it actually was caused by the vertical amplifier. If there were actually some distortion produced by the equipment under test, it would be impossible to distinguish it from the distortion of the scope.

Another cause of distortion that can be very misleading is an **impedance mismatch** between the audio generator or other signal source and the circuit being tested. For reliable results, the output of the generator must be properly matched to the test circuit. This is especially true when the square-wave section of the input generator is being used. Square waves are easily distorted by improper circuit conditions. That is why we use them for testing circuits. Care must be taken not to cause distortion of a square wave **outside** the circuit being checked.

Since a square wave is rich in harmonics, frequency and phase distortion are readily observed in such a wave. If a perfectly rectangular square wave is applied to a circuit, the shape of the output wave not only indicates **whether** trouble is present, it gives a clue as to **what** trouble is present.

While some audio generators produce both sine waves and square waves, there are many generators that produce only sine waves. Testing with these generators is not as informative or reliable as testing with square wave generators. A square wave generator is an especially important instrument for testing sound reproduction equipment, in which distortion must be held to the barest minimum. For this reason, the square-wave generator is becoming an increasingly common test instrument.

A square wave or any periodic (recurring at regular intervals) nonsinusoidal waveform can be shown to be the sum of: (1) a fundamental wave—that is, a sine wave of the same frequency as the original nonsinusoidal wave and (2) a series of harmonics—sine waves whose frequencies are integral multiples of the fundamental frequency.

All of the sine waves just mentioned are **components** of the original nonsinusoidal wave. With appropriate equipment or by (a difficult) mathematical analysis, we could either separate and identify its fundamental and harmonic com-

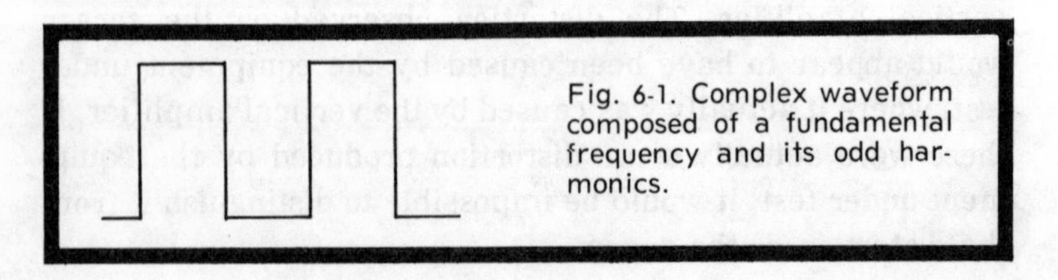

Fig. 6-1. Complex waveform composed of a fundamental frequency and its odd harmonics.

ponents, or we could combine an appropriate number of sine waves to produce a replica of the nonsinusoidal wave. The composition of a nonsinusoidal waveform is of interest because it helps us to understand the problems involved in generating, amplifying, processing and displaying complicated waveforms.

If we were to reproduce a nonsinusoidal waveform by adding sine-wave components together, each sine wave would have to have the correct amplitude, frequency, and phase to accurately reproduce the waveform.

Inasmuch as a nonsinusoidal wave is itself composed of two or more sine waves, we refer to a nonsinusoidal waveform as a **complex** waveform.

An example of a complex waveform is the sequence of rectangular pulses shown in Fig. 6-1. Although the pulses are actually rectangular, the waveform is referred to as a **square wave.** This waveform is constituted by a fundamental sine wave plus an infinite number of odd harmonics. In other words, the waveform in Fig. 6-1 is composed of the fundamental and only those harmonics whose frequencies are equal to the fundamental frequency multiplied by **odd** whole numbers: 3, 5, 7, ... The amplitude of any of the harmonics

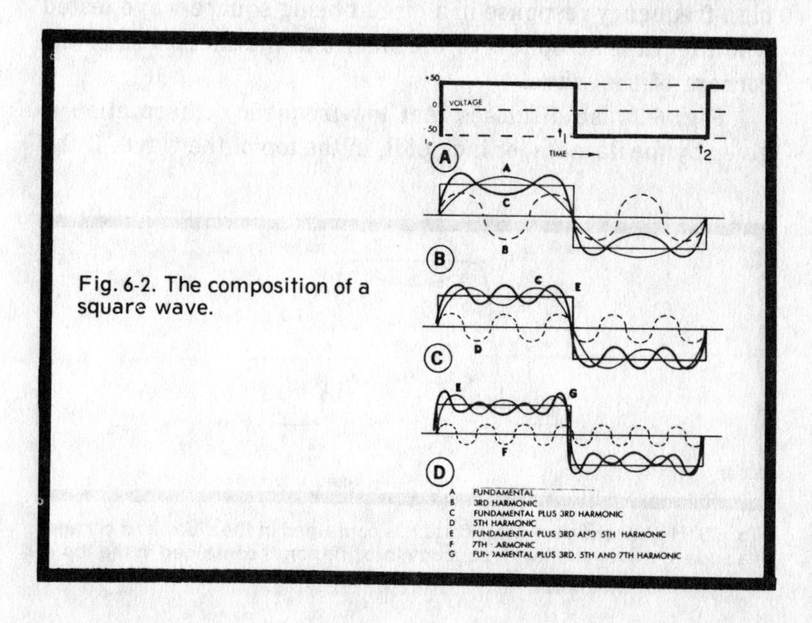

Fig. 6-2. The composition of a square wave.

comprising a square wave is inversely proportional to the frequency of the harmonic. The third harmonic is a third as strong as the fundamental, the fifth harmonic is a fifth as strong as the fundamental, and so on. How these harmonics combine with the harmonic to produce a square wave is depicted in Fig. 6-2.

Notice in Fig. 6-2C how the first two harmonics added to the fundamental provide an approximation of a square wave. As more harmonics are added, the resultant wave becomes a closer and closer approximation of the square wave as in Fig. 6-2D. Each additional harmonic causes the vertical edges of the resultant to be steeper and the corners to become sharper. A sufficiently high harmonic would fit the corners of the rectangles in Fig. 6-2 so closely that the resultant would appear to be square, or rectangular. But to produce a perfectly rectangular corner and a perfectly straight vertical edge would require an infinite number of harmonics. A perfectly rectangular square wave is impossible to attain. Fortunately, waves can be generated that are very close to this ideal.

As shown in Fig. 6-3, information concerning the amplitudes and phase relations of the high-frequency harmonics is contained in the edges and corners of the wave. A lack of high-frequency response in a circuit being square-wave tested will affect the steepness of the sides and the sharpness of the corners of the wave.

Fig. 6-3 also discloses that low-frequency information is given by the flatness, or lack of it, in the top of the wave. If the

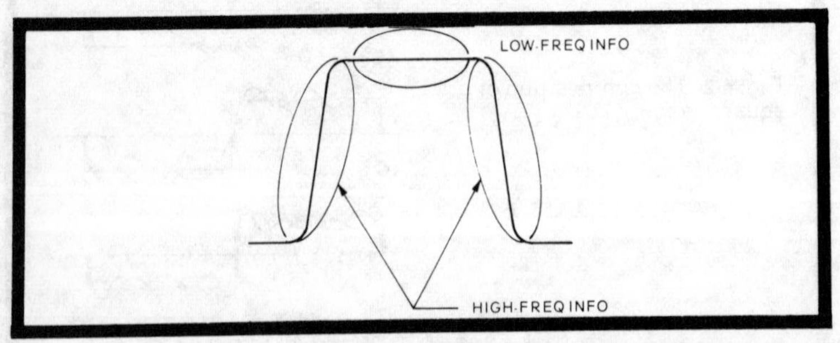

Fig. 6-3. High-frequency information is contained in the sides and corners of a square wave, and low-frequency information is contained in the top of the square wave.

low-frequency components do not have the proper amplitudes and phase relationships, this will be evidenced by a slope or curvature in the top of the wave.

Several common examples of distorted square waves are presented in Fig. 6-4. Fig. 6-4A is, of course, an undistorted square wave, shown for reference. This wave, at the output of an amplifier, indicates that no phase or frequency distortion has occurred in the amplifier.

When only the leading corners of a square wave are rounded, as in Fig. 6-4B, the equipment under test causes excessive phase delay at high frequencies. An indication of excessive phase delay at both high and low frequencies is when both the leading and trailing corners are rounded, Fig. 6-4C. Reduced height on trailing vertical edges, Fig. 6-4D, means the equipment tested has poor low-frequency response. Fig. 6-4E appears when the low-frequency components are accentuated. That both low- and mid-frequency response are below par is indicated by the waveform in Fig. 6-4F.

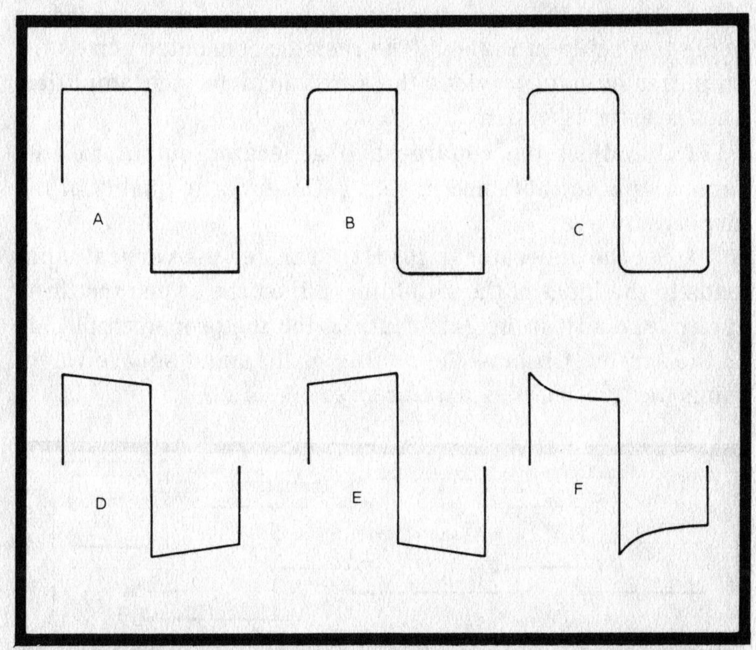

Fig. 6-4. A square wave and several examples of distorted square waves, discussed in the text.

As we said before, square waves are preferred over other waves for testing purposes because the square waves suggest the nature of a defect, rather than merely signaling its presence. By observing square-wave response, we can tell whether it is the high frequencies or the low frequencies that are affected in passing through or being amplified by a piece of equipment. Using sine waves and other waves, it is harder to distinguish between high- and low-frequency defects. In any case, if two linear devices give identical responses when square waves are passed through them, we can be sure they will give similar responses when other waves are passed through them.

Here is the specific procedure for determining what happens to an amplifier's characteristics with a square wave input signal.

1. Connect the square wave generator to the vertical input of the scope. Set the generator to 1 kHz, and adjust the scope controls for a display of about three cycles. Notice the quality of the square wave supplied by the audio generator.

2. Connect the generator, amplifier, and scope according to the test setup in Fig. 6-5. The resistor connected across the amplifier output, provides the rated load for the amplifier. This resistor is optional.

3. Readjust the square-wave generator output for the same amplitude obtained in step 1. Observe the quality of the displayed wave.

4. Set the generator to 100 Hz. Transfer the vertical input leads to the input of the amplifier, adjust the sweep for three cycles, and adjust the gain controls for the proper amplitude on the screen. Observe the quality of the input square wave, using the graticule as a reference.

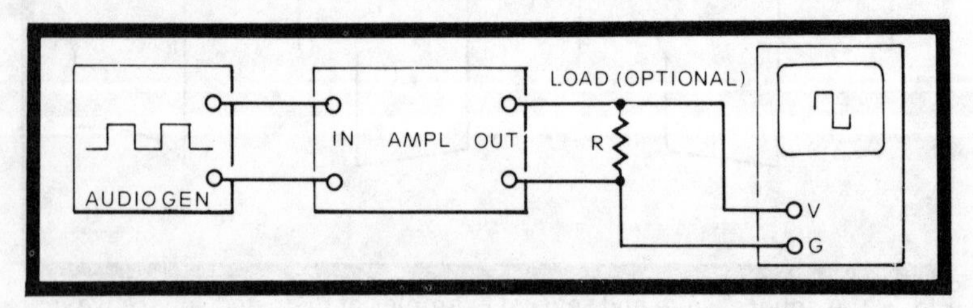

Fig. 6-5. The setup for square-wave testing an audio amplifier.

5. Shift the vertical leads back to the output of the amplifier, and readjust the gain controls for the proper display height. Observe the nature of the output waveform, using the graticule for reference. Compare the output waveform to the input waveform observed in step 4. Note especially the effect of the amplifier on the flat top portion of the signal waveform. This is the part of the square wave that carries the low-frequency information. If there is no deterioration of the flat portion of the square wave observable at the amplifier output, set the square wave generator to a lower frequency and repeat steps 4 and 5 until a noticeable deterioration in the flat portion does occur.

6. Repeat steps 4 and 5 for generator frequencies of 5 kHz and 10 kHz, in turn. Continue at higher frequencies until the output waveform has both sloping sides and rounded corners and begins to resemble a sine wave.

Using the procedure just described, you can graph the frequency response vs distortion of that amplifier.

WAVE SHAPING NETWORKS

When a square wave is applied to a **series RC** combination, the shape of the wave is altered, and the resultant shape depends on the relative sizes of the R and C components, and to a lesser extent on their absolute values. By varying the values of these components, the output pulse of the RC network may be shaped in **width** and in **amplitude** to suit different requirements. This can be demonstrated with a square-wave generator, a series RC network, and a scope.

Differentiated and Integrated Output Pulses

A capacitor charging in series with a resistor from a square-wave source will accumulate, or **integrate** a voltage over a period of time due to the current flowing in it. When a square-wave pulse is initially applied, the maximum current flows in the capacitor and resistor. The voltage across the capacitor, as shown in Fig. 6-6, rises exponentially during the flat-top portion of the square-wave pulse approaching maximum at a decreasing rate until the capacitor is fully charged. The initial accumulation of voltage on the capacitor is greatest because the charging current is greatest in the beginning.

The voltage across the resistor is in phase with the current that flows in the resistor, and thus is maximum at the outset. As the capacitor rapidly charges, it takes over the voltage of the resistor, and the voltage across the resistor decreases rapidly toward zero at a decreasing rate, while the voltage across the capacitor increases, as Fig. 6-6 shows. The sharp voltage pulse across the resistor is the **differentiated** pulse, easily recognized as **voltage spikes**. When the output is taken from across the capacitor, an **integrated** waveform consisting of rounded pulses results. Thus, the RC circuit may be either a differentiating circuit or an integrating circuit, depending on where the output is taken from.

As Fig. 6-6 shows, the leading edge of a square-wave input pulse initiates a positive-going output pulse, and the trailing edge of an input pulse initiates a negative-going output pulse—but, in this case, the integrated pulse across C cannot go below the arbitrary zero reference. In other designs, the integrated pulse might go from zero to a negative maximum—a little study will show that this will be the case if the present arbitrary zero reference on the input became −V, and the present top of the input wave became the zero reference.

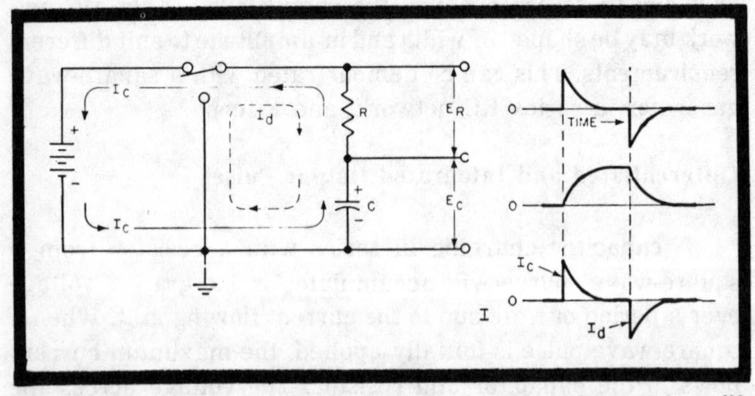

Fig. 6-6. When a square wave is applied to a series RC network, a **differentiated** waveform appears across the resistor and an **integrated** waveform appears across the capacitor.

RC Contents

How long it takes a capacitor to charge or discharge through a resistor depends on the values of C and R. The product of capacitance and resistance (in farads and ohms) is the time constant of the circuit in seconds. It takes about five time constants (5 RC) for a capacitor to fully charge or discharge. Since the width of the pulses derived from an RC network depends on the charging or discharging time of the capacitor, the width also depends on the time constant, or on R times C. In such a network, such a capacitor is called the **timing capacitor**.

Using the Pulses

Since the differentiated pulse is a sharp spike having a very brief duration at its peak, it is very suitable for triggering or sync purposes. By varying the capacitance of the wave-shaping network, we can vary the width of the differentiated sync pulse. Another factor in TV sync circuits is the **off** time of the input square wave, the time **between** the square-wave pulses. If successive sync pulses are to be distinct from one another instead of running together, the **off** time must be sufficient for the capacitor to fully discharge before the next square-wave pulse begins. Thus, TV horizontal sync pulses have an **on** time of about 5 usec, and an **off** time more than ten times as long (64 usec, roughly). In contrast, the serrated **vertical pulses** are on much longer than they are off, so that they integrate, or combine. Energy accumulates from one vertical pulse to the next, with only a small loss during the off time.

Principles of Wave-Shaping Networks

1. Build the circuit shown in Fig. 6-7, using a 100 pF capacitor and a 10K resistor across the output of the audio generator. Set the generator to 1 kHz.

2. Measuring across the generator output, adjust horizontal sweep of the scope for a display of three cycles, and note the quality and amplitude of the square wave.

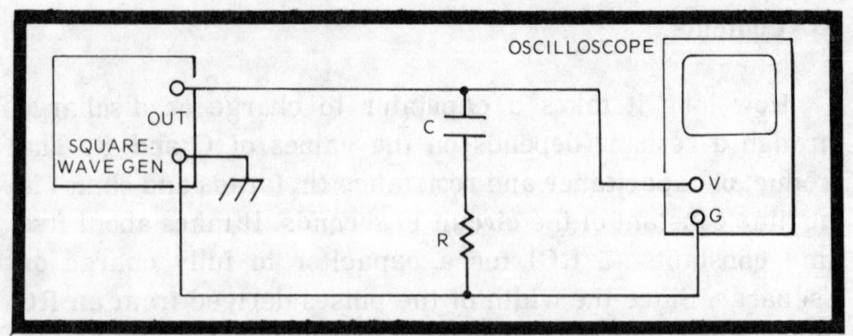

Fig. 6-7. An arrangement for observing differentiated and integrated square waves.

3. Change the generator setting to 10 kHz, and again note the quality and amplitude of the square wave.

4. Now transfer the vertical input leads of the scope to the resistor, and return the generator to 1 kHz. Draw the resultant pulses on a sheet of paper and note the relationship between the on time and the off time.

5. Shift the vertical input leads to the capacitor, and sketch the displayed pulse alongside the one obtained in step 4, for comparison.

6. Reset the generator for a frequency of 10 kHz, and observe, in turn, the pulses across R and C. Sketch these pulses and compare them to the ones obtained at 1 kHz.

7. Change R to 100K and C to 0.001 uF. Observe and record the waveforms across R and C, in turn, at a frequency of 10 kHz.

8. Set the generator for 1 kHz, and observe and record the waveforms.

9. Change C to 0.01 uF. Leave R at 100K. Observe the waveform across the resistor, and then across the capacitor. Do this for frequencies of 1 kHz and 10 kHz, and sketch the waveforms obtained.

10. Study your sketches and review the theory vs your results.

ELECTRONIC SWITCH FOR DUAL-TRACE DISPLAYS

Often it is desirable to observe two signals at the same time. It might be desired to observe and compare, for example, the input and output waveforms of an amplifier simultaneously on the same crt screen. Dual-trace scopes, to be discussed later, have a built-in capability for such observations, but dual-trace observations can be made with an ordinary single-trace scope, with the aid of an electronic switch.

The electronic switch switches the vertical input of the scope between the two signal sources at a very fast rate, perhaps 100,000 times per second. Basically the electronic switch is a balanced multivibrator in which each collector (or plate) circuit alternately connects to the vertical input of the scope. Since each collector (or plate) conducts, or is on, during alternate half-cycles, any signal applied to either transistor (or tube) will be displayed on the screen only during alternate half-cycles. Although only one signal is displayed at a time, the repetition frequency of the sweep, the persistence of vision, and the persistence of the crt phosphor make both signals appear simultaneously.

Since the two signals displayed simultaneously by means of an electronic switch are sampled alternately at a 100 kHz rate, the traces are chopped up into a series of dots. Thus, the mode of operation of an electronic switch is the **chopped** mode. You will meet up with this term again in the discussion of dual-trace scopes.

Fig. 6-8 shows a dual-trace display of the collector and base waveforms of an astable multivibrator, as they might be

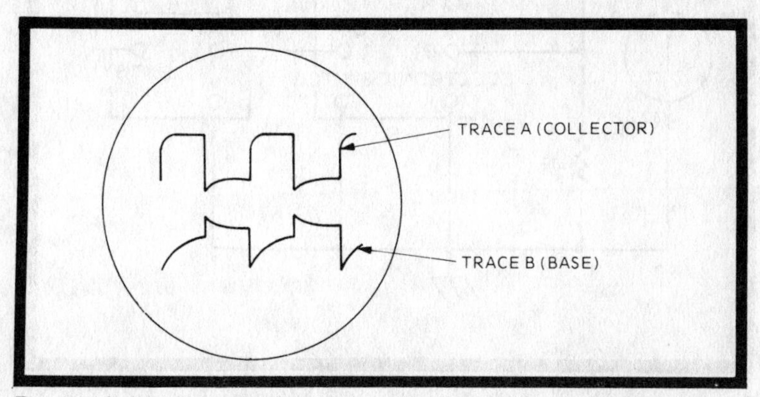

Fig. 6-8. A dual trace display of the collector and base waveforms of an astable multivibrator.

viewed by means of an electronic switch. The timing relationships can be readily observed. Phase relations are also readily observed in a dual-trace display, as you can see by following the procedure given next.

1. Build the test circuit in Fig. 6-9. For detailed instructions on operating the electronic switch, refer to the manufacturer's instructions. Note that the output of the square wave generator is connected to input A of the electronic switch, and also to the **external sync** terminal of the oscilloscope. Input B is connected to show the differentiated square wave that will appear across the resistor. The output of the electronic switch, which consists of the square wave at input A and the differentiated square wave at input B, is coupled to the vertical input of the oscilloscope.

2. Adjust the square-wave generator for a 1 kHz output of 10V p-p.

3. Adjust the controls of the electronic switch and of the scope until you obtain a stable display on the screen, with

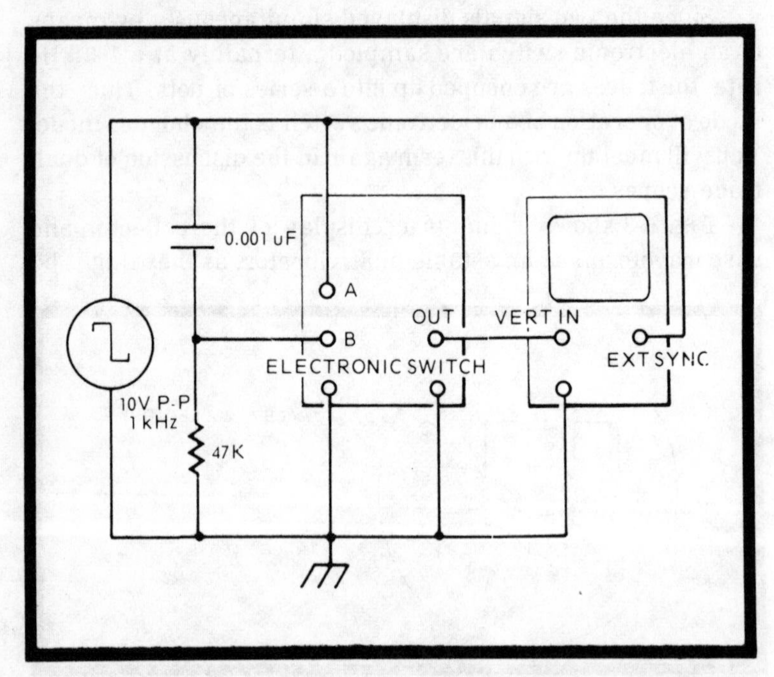

Fig. 6-9. An electronic switch connected to permit simultaneous observation of the waveforms across R and C of a series RC circuit.

three complete square waves on the upper portion of the screen and, below the square waves, the waveform of the voltage across R. Draw the waveforms on a sheet of paper, showing the proper phase relationship.

4. Transpose R and C so that input B is connected to show the integrated square wave that appears across C. Draw the waveform of the voltage across C below the waveforms drawn in step 3, showing the proper phase relationship to the square wave. Your drawing should be similar to the one in Fig. 6-6.

INSTANTANEOUS-VOLTAGE MEASUREMENT

One of the advantages the scope has over ordinary meters is its ability to display instantaneous values of voltage. A peak-reading voltmeter gives the value of one instantaneous voltage, but an infinite number of instantaneous voltages are represented in a waveform on the scope screen. Before one can measure instantaneous voltages with a scope, it is necessary to calibrate the scope for instantaneous voltage measurements. You calibrate and make measurements as follows.

1. Set up the scope to display a 60 Hz voltage of 3 cm or 3 inches p-p. Arbitrarily call this a display of 3 **volts** p-p. Adjust the scope until one complete cycle is displayed over a path of about 4 cm, if the graticule of the scope is calibrated in cm, or 4 inches, if the graticule is calibrated in inches. Refer to Fig. 6-10.

2. Since one cycle contains 360 deg and the displayed cycle covers 4 cm or 4 inches, each cm or inch represents 90 deg, and each half-division of the graticule represents 45 deg. Thus, it is possible to calibrate the graticule at 45 deg intervals, as shown by Fig. 6-10.

3. Determine the instantaneous voltage of the sine wave at the following angles: 0 deg, 36 deg, 135 deg, 225 deg, 270 deg, 300 deg, and 315 deg. Do this by measuring the length of a perpendicular extending from the vertical axis at the proper angle to the sine wave. Since 3 increments equal 3 volts, each cm or inch equals 1 volt. Immediately, you can see that the voltage at 45 deg is about 1.05 volts.

4. Using the instantaneous voltage at 45 deg, e, calculate the peak voltage, E_{max}, using the formula

$$e = E_{max} \sin \theta$$

5. Compute the period of the waveform, i.e., the time taken up by one cycle. Calibrate the graticule in time, noting that 45 deg corresponds to 2.09 msec, as shown in Fig. 6-10.

6. Determine the instantaneous voltage of the waveform at the following times: 4.17 msec, 5 msec, 12.5 msec, and 11.1 msec.

7. Check your results in step 6 using the following formula:

$$e = E_{max} \text{ Sin } \omega t$$

where e is the instantaneous value, E_{max} is the peak value, and μ stands for $2\pi f$.

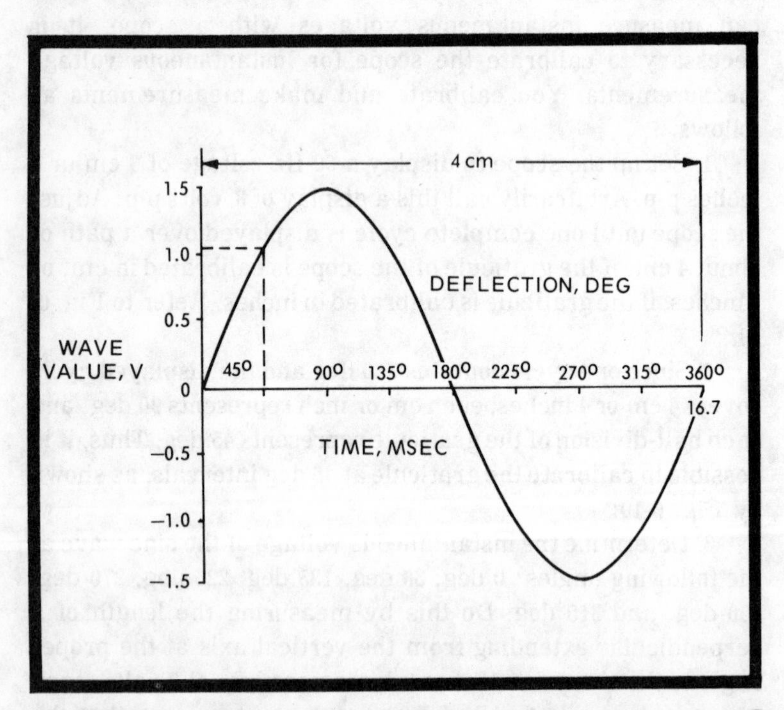

Fig. 6-10. Illustrating the calibration of an oscilloscope graticule for the measurement of instantaneous voltages.

MEASURING VOLTAGES IN A CIRCUIT

Since the beam of the scope is deflected by the voltages on the deflection plates, the scope responds to voltages, and it is basically a **voltmeter**. (If the scope crt employed magnetic deflection, as does the crt of a TV set, then the electron beam would be deflected by the current in the deflection coils and the scope would be basically an **ammeter**.) Unlike most voltmeters, the scope does not give a direct reading of rms voltages. It does give a direct reading of peak and peak-to-peak voltages for any waveform. For sine waves, the rms value of a voltage may be calculated according to the formula:

$$E_{rms} = 0.707\,E_{max}$$

The principles just explained may be illustrated by following the procedure outlined below.

1. Build the circuit of Fig. 6-11, using the 6.3V secondary voltage of a filament transformer as the source of emf. Note that 6.3V rms corresponds to 18V p-p.

2. Using the procedure outlined earlier, calibrate the scope for 5 volts per graticule division.

3. Use the scope to measure the peak voltages around the circuit, and see whether they approximate the voltage source. Peak voltages, like other voltages, obey Kirchhoff's voltage law, which states that the sum of the voltage drops around a circuit is equal to the source voltage.

4. With a VTVM (TVM), measure the voltage drops, both peak and rms. Compare the peak voltages to those measured with the scope.

5. Calculate the rms values of the peak voltages read with the oscilloscope, using the formula $E_{rms} = 0.707\,E_p$. Compare these calculated rms values with the ones obtained with the VTVM (or TVM).

POWER LINE HUM VOLTAGE PICKUP

Like any current-carrying wire, the wiring in a house acts as an antenna. The field set up around the electrical service

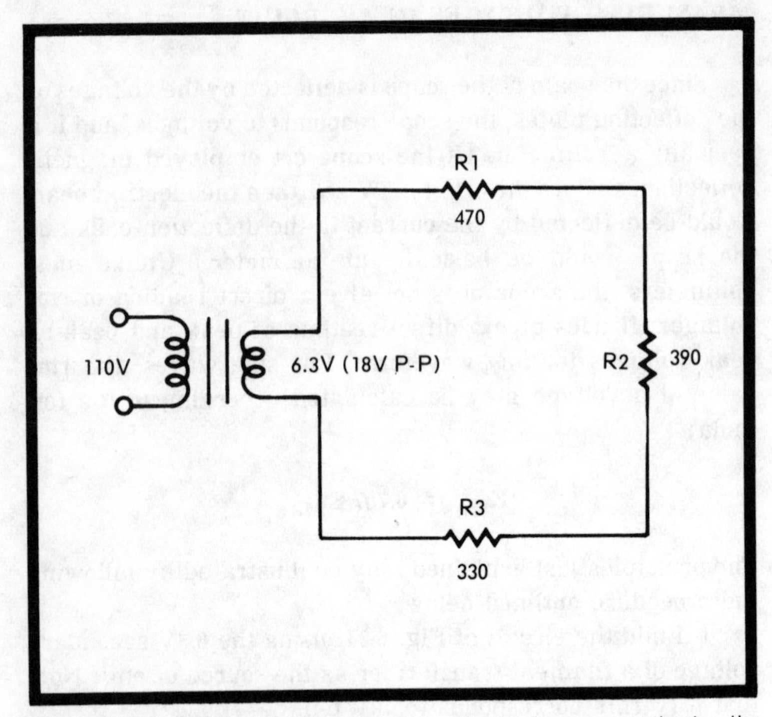

Fig. 6-11. The voltages around a series circuit may be measured using the procedure described in the text.

wiring is referred to as a **stray field**, since it represents electrical energy that has strayed from the wiring. Using properly shielded test leads will usually eliminate the problem of stray-field voltages, but it is important to understand how this type of interference can occur. It is also important to understand the effect of 60 Hz power line emi on a signal. Occasionally, a leaky power-supply filter capacitor will permit 60 Hz hum to enter a signal path, and the scope can detect and localize the source of this type of distortion.

The following procedure will show how stray voltages may be picked up, and how they will affect a square-wave test signal.

1. With the scope set up to display a 60 Hz sine wave, place the vertical input leads in the vicinity of a wire carrying 60 Hz current obtained from the power line. A distorted sine wave should result.

2. Connect the vertical input leads of the scope across a 100K resistor, and note the display that results.

Discussion. The display obtained in step 1 illustrates stray pickup. It is made possible by the fact that the vertical input circuit of the scope has a high impedance and a stray field is a high-impedance source. The open test leads were coupled by means of a small stray capacitance to the power-carrying wire in step 1. Since capacitive reactance decreases as frequency increases, the harmonics in the waveform were accentuated, and they showed up as kinks in the sine wave.

When the vertical input impedance was reduced to 100K in step 2, the stray-field pattern disappeared. This is because an extreme impedance mismatch then existed between the stray field and the vertical input. The input impedance was too low to couple into the high-impedance stray-field voltage.

3. Connect a square wave generator to the vertical input of the scope, and set the generator and scope to display two or three cycles of a 60 Hz square wave.

4. Remove the lead between the generator ground and the scope ground, and observe how the square wave is smeared, or thickened. The smearing or thickening of the trace illustrates the effect of 60 Hz hum voltage on a square-wave pattern.

USING A SCOPE AS A NULL INDICATOR

One common way of measuring frequency involves the use of the superheterodyne principle. The unknown frequency to be measured is mixed with a reference signal from a signal generator. The mixing of the two signals provides an i-f that is the difference between them. When the source of the known frequency is adjusted so that the known and unknown frequencies are the same, the i-f frequency is zero. The frequency at which this occurs is known as the **zero-beat** frequency or **null point**. And the scope is an effective null indicator, providing a visual indication.

Fig. 6-12 illustrates the system used in checking frequency by heterodyning a known frequency signal (F) with an unknown frequency signal (F_X). The two signals are fed to a nonlinear device such as a diode detector. (A demod probe may serve as the nonlinear detector.) The signals beat together in the detector and the output of the detector contains the sum frequency F plus F_X and, of course, the difference frequency F minus F_X. As frequency F is adjusted near F_X, the difference frequency is brought into the af range, until it equals F_X, and the difference frequency is zero.

In Fig. 6-12 the difference frequency is amplified by the null amplifier and fed to the indicator, a scope whose sweep frequency is set to 30 Hz. This gives a visual indication of the approach to zero-beat. Fig. 6-13(a) shows the pattern on the screen as F minus F_X falls into the audio range. Since the sweep is set to 30 Hz and 5 cycles are observed on the screen in Fig. 6-13(a), the display indicates that F minus F_X equals 150 Hz. As F minus F_X is brought progressively closer to zero,

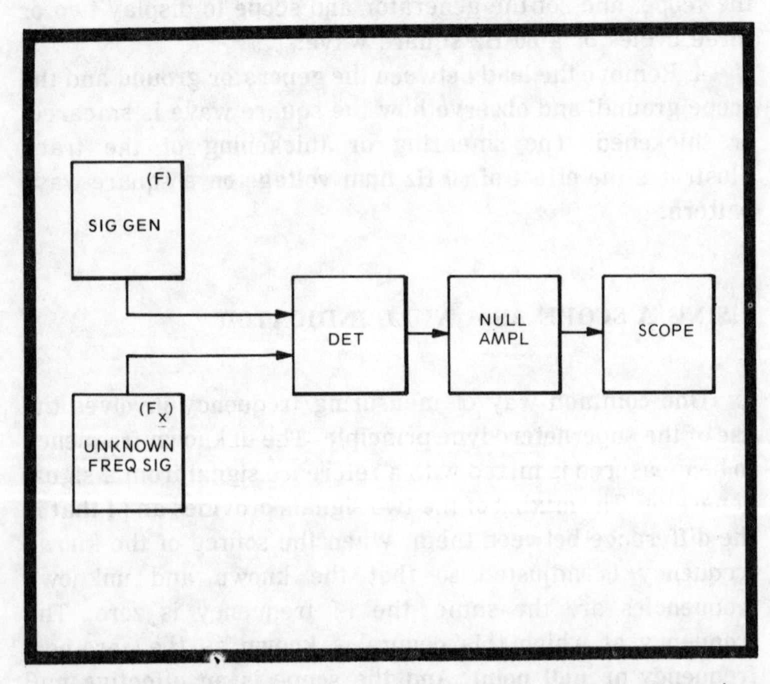

Fig. 6-12. The system used in checking frequency by the heterodyne method.

fewer and fewer cycles appear in the display. Finally, at zero beat, the frequencies are equal, and no cycles of difference are displayed as in Fig. 6-13(b). This display is an ideal, and if it is realized at all, it will be realized only momentarily, due to the drift of the signal generator.

We have assumed that the signal generator was originally set to a frequency higher than F_X, and that to obtain zero-beat the generator frequency F was reduced to the point where F equaled F_X. If the frequency of the generator were then reduced still further, until it was 150 Hz **below** F_X, the pattern in Fig. 6-13(c) would appear. Notice that this pattern is, of course, identical to the one in Fig. 6-13(a). In between the two frequency settings of the signal generator at which the display of Fig. 6-13(a) and (b) is obtained, lies the null point. The null point is the point on **either** side of which there is an audio-frequency difference, as the frequency control of the signal generator is rocked on either side of F_X.

One shortcoming of this method of frequency determination is that it is possible to obtain a zero-beat from certain harmonic combinations of the two input signals. The null indication having the highest amplitude is the one desired. It is the one that indicates the zero-beat of the two fundamentals. A procedure for demonstrating the use of the scope as a null indicator follows.

1. Assemble the circuit of Fig. 6-14. E1 is a sine-wave generator set to 100 kHz, and adjusted for 1V p-p output. E2 is set for 120 kHz and 1V p-p.

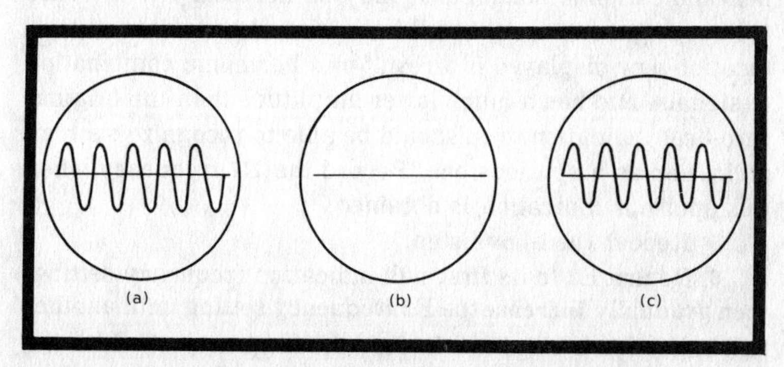

Fig. 6-13. Heterodyne-beat patterns on an oscilloscope.

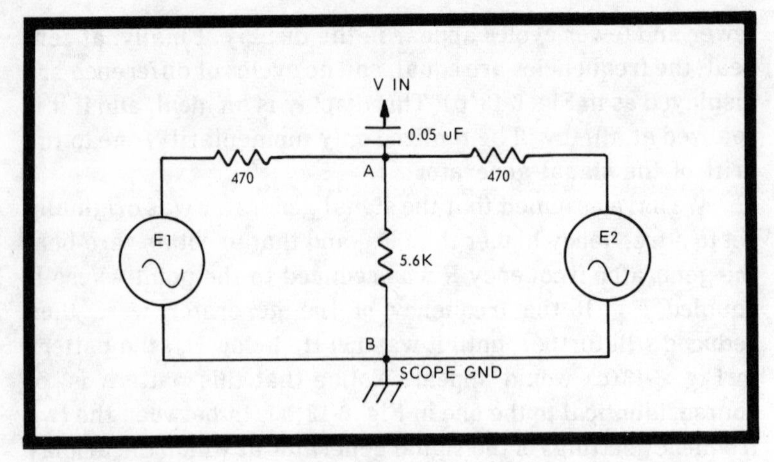

Fig. 6-14. A circuit for obtaining null indications on a scope.

2. Adjust the horizontal sweep to 30 Hz and set the sync selector switch of the scope to the "internal" position. Connect a demod probe to the vertical terminals of the scope. Connect the demod probe to point A in the test circuit, and connect the oscilloscope ground to point B. Connect a 0.05 uF capacitor across the vertical input of the scope. Adjust the frequency of E2 and the vertical gain of the scope until you obtain a display similar to the one in Fig. 6-13(a) or (c).

3. Gradually reduce the frequency of E2 until zero-beat is reached. Write the frequency at which zero-beat occurs. If both generators are properly calibrated, E2 should now be at 100 kHz.

4. Slowly reduce the frequency of E2 until another zero-beat indication is obtained. It may be necessary to increase the scope gain to see the null indication. At any rate, the indication now displayed is a result of a harmonic combination of signals, and has a much lower amplitude than the original zero-beat indication. You should be able to recognize such an indication as a spurious one. Record the E2 frequency when this spurious indication is obtained.

5. Repeat the above step.

6. Return E2 to its first null-indication frequency setting, then gradually **increase** the E2 frequency setting until another null indication is obtained. Record the frequency at which the indication is obtained.

7. Gradually increase the E2 frequency some more, until the next null indication is obtained. Note that the spurious null indications obtained in steps 5, 6, and 7 have much less vertical amplitude than the original null indication obtained in step 3.

OBSERVING AUDIO SIGNALS

Electrical variations representing speech and music are called **audio signals**. They are found in the circuits of radios, TV's, phonographs, telephones, radio transmitters, and public-address systems. The scope can display such signals; it is a very useful device for checking the circuits just mentioned.

Audio signals are complex waves, consisting of combinations of sine waves of different frequencies and amplitudes. They have the general appearance of the waveforms in Fig. 6-15. Audio signals representing speech or music will not produce stationary displays as did the simple, constant signals of the sine-wave generator or the power line. The reason is that as music or speech is sent into an amplifier, the continual variations in the loudness of the speech or music will cause the amplitude of the audio signal to vary continually. Also, the various frequencies produced in speech or music are constantly changing. This also will cause the shape of an audio waveform to change continually.

Audio signals are observed on a scope during signal-tracing procedures, and their observation is of considerable value during testing and troubleshooting of audio equipment. Since the audio signals are changing continually in amplitude and wave shape, we do not try to measure their amplitude; nor are we very concerned with their shape during signal tracing. Our main purpose in signal tracing is to observe whether a signal is present or absent, and at what point in a circuit it becomes absent.

For effective use of the scope in testing audio circuits, it is important to become familiar with the appearance of audio signals on a crt screen. The following procedure will demonstrate the observation of audio signals and aid in recognition of their general wave shape in the future.

1. Connect a crystal microphone to the vertical input of the scope, as in Fig. 6-15(a). Set the horizontal control to the 1 kHz-10 kHz range, and observe the waveforms produced on the screen when you whistle or speak into the microphone. Adjust the scope controls as necessary to obtain a clearly defined, well centered display of sufficient amplitude.

Discussion. Unless you are using shielded test leads, the test leads may pick up stray ac voltages, producing a sine wave on the screen. For our purposes here, it is unnecessary to eliminate the stray field pickup. Actually, the appearance of such a sine wave serves to demonstrate a principle—the modulation of a fixed sine-wave signal by audio signals. The

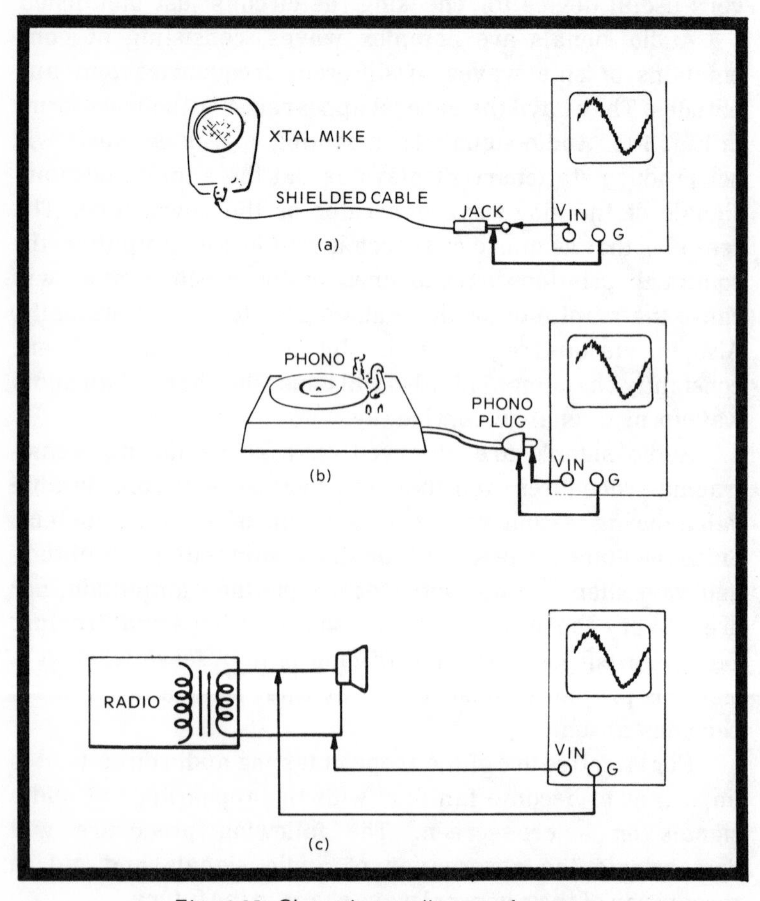

Fig. 6-15. Observing audio waveforms.

audio signals will develop around the 60 Hz sine wave that appears due to stray pickup. It may be possible to greatly reduce (partially cancel) the stray pickup by twisting the vertical input leads together. Try it.

2. Vary the sweep vernier control and notice the effect on the audio signal waveform. Vary the setting of the horizontal sweep selector switch and observe the effect on the waveform.

3. Attach the vertical input leads of the scope to the output plug of a phonograph pickup plug, as in Fig. 6-15(b). Play a phonograph record and note how the variations in the audio signal are displayed on the screen. Vary the vernier and horizontal sweep controls, and note the effect on the observed waveforms.

4. Attach the vertical leads of the scope to the voice coil leads of a radio or phonograph loudspeaker, as in Fig. 6-15(c), and observe the changes in the amplitude and wave shape of the various audio signals displayed on the screen.

BASIC SIGNAL TRACING

Signal tracing a piece of electronic equipment consists of tracing a signal present in the piece of equipment through the various stages to find where the signal is **lost**. Thus, when a piece of equipment is inoperative, we can use this technique to localize the trouble. Once we have found the stage where the signal is no longer present, we can confine our testing to that particular circuit without having to hunt for defective components in all the other circuits.

In general, when signal-tracing the phonograph, radio, or audio sections of a TV, we start at the audio **input** of the device. This is commonly the volume control, as shown at test point 8 in Fig. 6-16. The audio input at this point could be from the detector of a receiver, or from a microphone or a phonograph pickup. If we found audio here, we would continue the signal tracing by placing the scope lead at test point 7 (TP 7), the base of the transistor. The signal should still be present here, and if it is, we would make our next check at TP 6, the collector of the transistor. Here again we would expect a signal to be present. We would also expect the signal to be somewhat larger at this point, reflecting the gain of this common-emitter stage; but we would not attempt to make any actual measurements of the gain. To repeat: The primary purpose of signal tracing is locating a signal, not measuring its amplitude nor analyzing its shape.

In signal tracing the amplifier of Fig. 6-16, we would continue to work through it from left to right, and at the place where no signal is observed, we have isolated the defective circuit. Of course, if there were no audio at VOL, we'd work

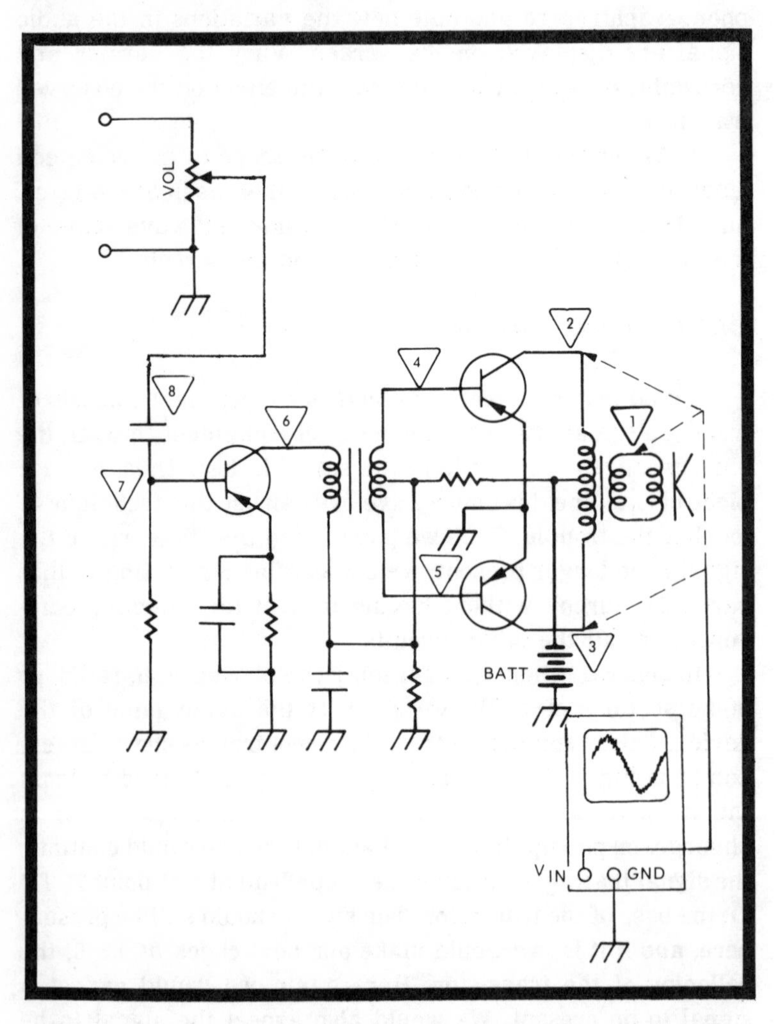

Fig. 6-16. Signal tracing a transistorized audio amplifier with a scope.

forward (left) toward the preamps (audio) and rf sections, changing scope sweep accordingly.

We could just as well start at the right of the circuit of Fig. 6-16, and work back toward the input. We could look for the signal at the speaker voice coil, TP 1, and if we found no signal there, move the probe back to TP 2 for the next check. If a signal were found at TP 2, then we would have isolated the trouble between TP 1 and TP 2. Or, if no signal were found, we would move on to TP 3, and continue in this manner until we found a signal.

The output stage of the amplifier of Fig. 6-16 is a push-pull circuit. Such circuits are frequently employed in high-quality phonographs and other audio systems. The use of dual transistors in push-pull results in increased audio output power and low distortion. Sometimes audio output may be obtained from a push-pull stage even though one or the other of the push-pull transistors has burned out. Thus, some signal might be present at TP 1 even though one of the dual transistors was burned out. The scope would reveal this defect. If one of the transistors were burned out, the checks at TP 2 and TP 3 would disclose which one it was. If a defect is found at TP 2 or TP 3, you should also check at TP 5 and TP 6. It might be that half of the secondary of the interstage transformer has burned out, removing the signal from the base of one of the push-pull transistors. With no signal at the base, one could not expect to find any signal at the collector of the transistor.

AMPLITUDE MODULATION CHECKS

In an amplitude modulation (AM) radio transmitter, an audio frequency electrical signal, not capable of propagating through space, is superimposed on a higher frequency rf (radio frequency) carrier wave, which is then radiated from an antenna. As Fig. 6-17 shows, when an audio waveform (a) is superimposed on and modulates an rf carrier (b) in the AM process, a modulated carrier that is a faithful replica of the audio wave results (c). Although a sinusoidal audio-modulating waveform is shown in Fig. 6-17A, the modulating waveform is usually a complex wave. As a result, the modulated carrier waveform, called the **modulation envelope**, will also have a complex nature, as depicted in Fig. 6-18 (b).

During intervals between words and musical passages, the unmodulated carrier wave at constant amplitude can be observed in the modulation envelope. This is a sine wave, as can be seen in Fig. 6-17B. When the unmodulated rf carrier is displayed, there is said to be zero percent modulation. On the other hand, when the carrier amplitude is reduced to zero, there is said to be 100 percent modulation. Under this condition, the carrier will be cut off altogether, and harmonic radiations and a distorted signal will result. To prevent this, modulation percentage is constantly monitored at a broadcast station, and kept well below 100 percent, except on extreme peaks of the audio signal. Special modulation monitors approved by the FCC are used. These give a meter indication of the percentage of modulation. Often, the modulation envelope is monitored as well.

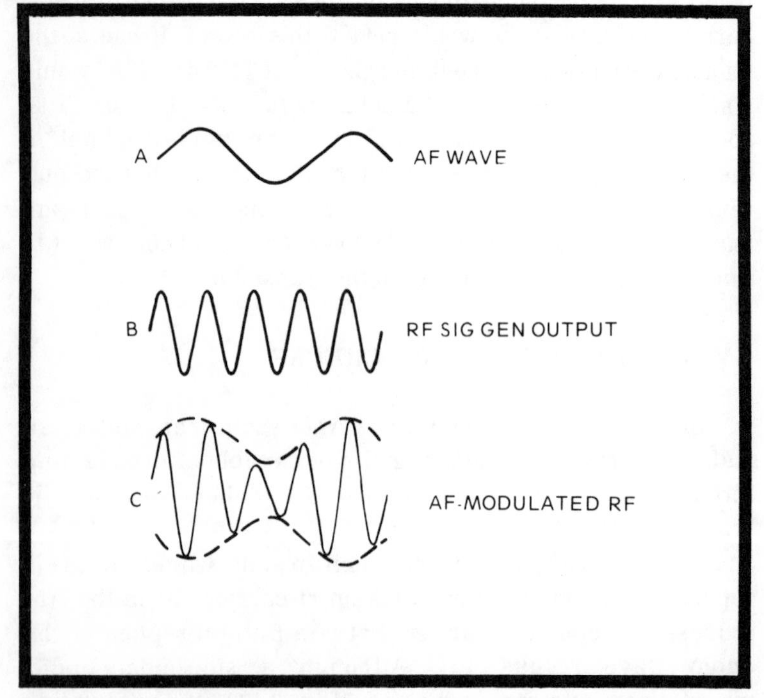

Fig. 6-17. An audio frequency may be changed to a radio frequency by the process of amplitude modulation.

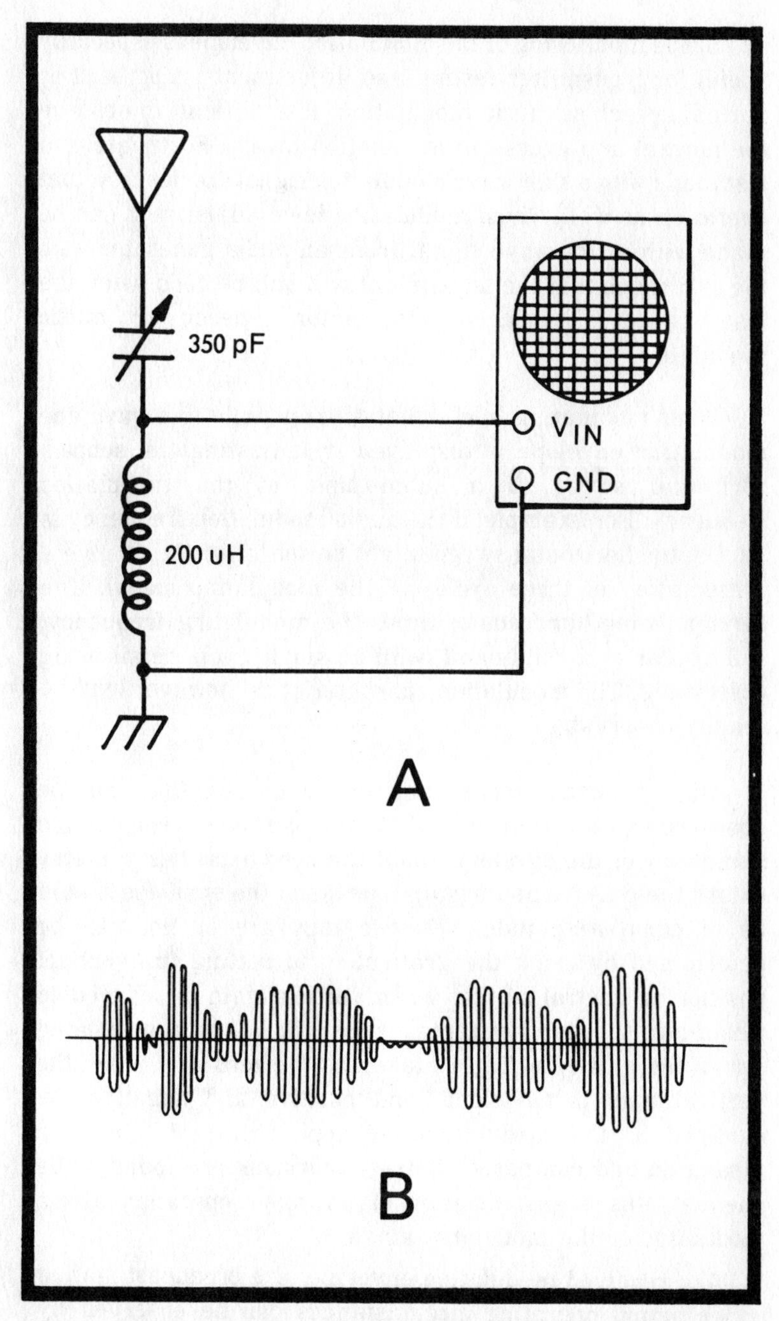

Fig. 6-18. (a) Setup for obtaining AM radio waveforms. (b) Typical AM radio modulation envelope.

Scope monitoring of the modulation envelope is especially useful for transmitter testing and adjustment purposes. For normal speech or music modulation, it is difficult to observe the normal and excessive modulation levels. Fortunately, a test made with a sine-wave modulation signal applies to actual operation as well. Thus, modulation level adjustment can be made with a sine-wave signal from an audio generator with the assurance that the adjustment will still be good when the test signal is replaced with actual speech or music modulation.

When modulating an rf signal with an audio sine wave, the modulation envelope is displayed by adjusting the scope's horizontal sweep to a submultiple of the modulation frequency. For example, if the audio modulation frequency is 300 Hz, the horizontal sweep might be set to 100 Hz, to give a presentation of three cycles of the modulation signal. The carrier, being hundreds of times the modulating frequency, will appear as a solid band, with no single-cycle detail being observable. The modulation characteristics, however, will be readily observable.

One important modulation characteristic that can be observed in a scope display of the modulation envelope is the symmetry of the envelope about the zero axis. If symmetry exists, the positive and negative peaks of the envelope should be of equal amplitude. Whether they are or not can be determined by using the graticule and setting the vertical positioning control so that when vertical gain is reduced to minimum, the thin horizontal trace will fall on a graduated line of the graticule. This is taken as the zero axis. Then the vertical gain is advanced, and positive and negative excursions of the modulation envelope from this line are measured and compared. If the excursions are found to be unequal, this is an indication of improper operation of the modulator or the modulated stage.

The received modulation envelope of a broadcast station under actual operating circumstances can be observed by connecting the test setup of Fig. 6-18. An ordinary broadcast coil and capacitor are used in the test setup to tune in the desired station. The horizontal sweep is set to the 1-10 kHz range and the display is centered about the horizontal axis by collasping the display vertically and adjusting the vertical positioning control. The vertical gain is then adjusted for proper pattern height. Of course, it will not be possible to lock in the display on the screen, since the audio signal being observed is varying constantly in frequency. The approximate percentage of modulation M can be determined from

$$M = \frac{100\,(a - b)}{a + b}$$

where a is the maximum amplitude of the envelope and **b** is the minimum amplitude of the envelope.

An even more common scope check of modulation percentage utilizes the test setup of Fig. 6-19. In this setup, the crt beam is swept across the screen horizontally by the audio modulating frequency instead of the scope time-base

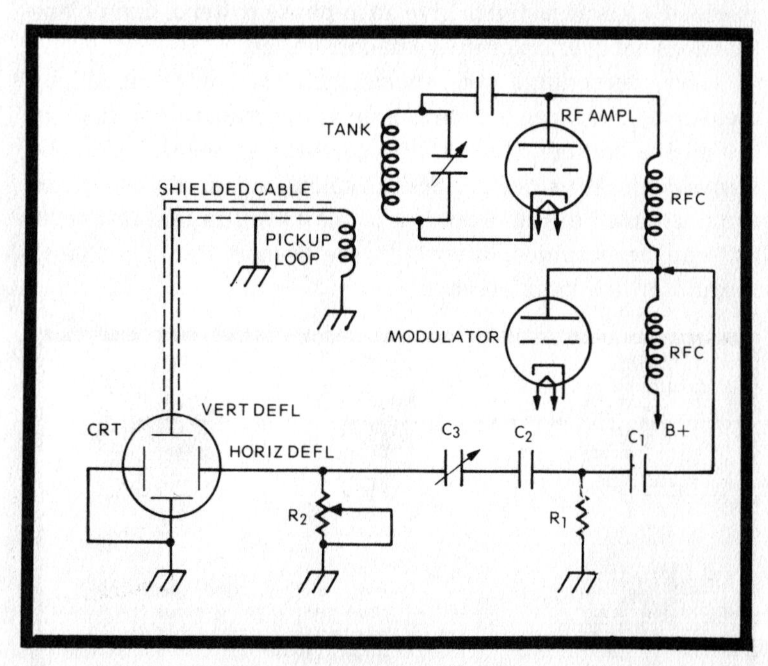

Fig. 6-19. A test circuit for obtaining trapezoidal AM patterns.

generator. Also, the vertical amplifier of the scope is bypassed, and the modulated rf signal is applied directly to the vertical deflection plates. Almost any scope, regardless of quality, can be used for this test, since the only part of the scope used is the crt, and the crt has a very wide frequency response. Of course, the scope must have provisions for a direct connection to the vertical deflection plates. (If the scope used is a high-quality instrument having a wideband response, it is not necessary to bypass the vertical amplifier; the modulated rf signal can be loosely coupled to the vertical amplifier input.)

In Fig. 6-19, C1 and C2 are 0.005 uF capacitors rated to withstand the high modulator plate voltage. C3 is a trimmer. R1 is chosen to provide a suitable horizontal deflection, and R2 is a 50K potentiometer.

In making the modulation test, an audio generator may be used to replace the speech or music modulation of the transmitter. The pickup loop is placed so as to provide proper vertical deflection, and the lead from the loop to the crt is shielded. C3 is adjusted to give an in-phase pattern, that is, one that does not overlap or give a double image.

The percentage of modulation is determined by measuring the larger vertical side of the trapezoidal display, a, and the smaller side, b, and calculating according to the formula for M as before. Some typical modulation patterns are illustrated in Fig. 6-20. If a display such as that in Fig. 6-20C can be obtained, then both the transmitter signal and modulator are satisfactory.

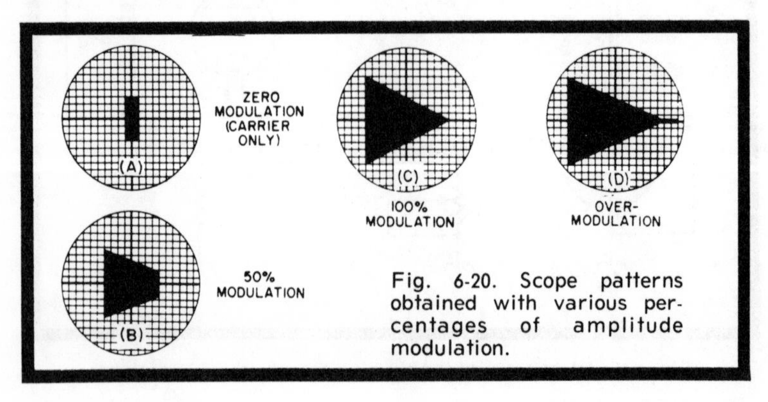

Fig. 6-20. Scope patterns obtained with various percentages of amplitude modulation.

One advantage of this method of measuring modulation is that the waveshape of the display is independent of the waveshape of the modulating signal. Thus, a good indication of percentage of modulation, even during normal program transmission, can be obtained.

AMPLIFIER GAIN AND FREQUENCY RESPONSE

An audio generator and a scope can be used to measure the gain and frequency response of an audio amplifier. Fig. 6-21 shows the test setup for making these measurements.

Gain

To measure the gain of an amplifier at any specific frequency, a measured sine-wave voltage (ein) is applied to the input of the amplifier, and the output signal of the amplifier (eout) is then measured. The gain of the amplifier at the applied factor is given by the formula

$$G = \frac{eout}{ein}$$

In order to make a meaningful determination of gain, take care not to overload the amplifier and cause distortion. Since

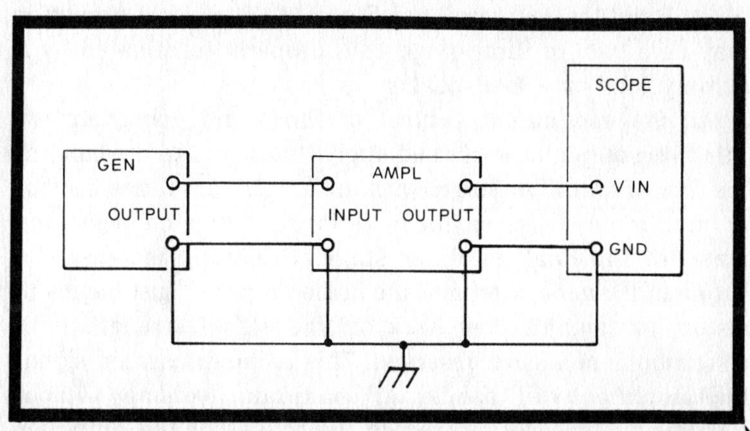

Fig. 6-21. Equipment connected to check an amplifier's gain and its frequency response.

distortion can be readily observed, a scope measurement of gain can be very reliable. In measuring the input and output signals with a scope, it is usually best to make peak-to-peak measurments.

The gain of an amplifier will remain constant over the **linear** portion of its collector (or plate) characteristic. By decreasing and then increasing the signal applied to an amplifier while observing the output waveforms, it is possible to determine the range of signal amplitudes over which the gain of the amplifier is constant.

Frequency Response

The frequency response of an amplifier can be determined by applying a fixed-amplitude sine wave at various frequencies to the input of the amplifier, measuring the output, and plotting a graph of output voltage vs frequency. The level of the input signal must be kept below the level at which distortion occurs.

Another method of checking the frequency response of an amplifier is to measure the **gain** of the amplifier at each of a number of different frequencies in the range of the amplifier, and to plot a graph of **gain** vs **frequency**. This graph will have the same shape as a graph of output voltage vs frequency.

Here is a detailed plan for checking the gain and frequency response of an audio amplifier.

1. Build the test circuit of Fig. 6-21. The audio amplifier may be a two- or three-stage unit, properly terminated by a dummy load or a loudspeaker.

2. Set the output control of the audio generator for **minimum** output at 1 kHz and apply this signal to the input of the first af amplifier stage, with the amplifier volume control at **maximum**. Observe the waveform at the collector (or plate) of the final amplifier stage. Gradually increase the output of the generator until the audio amplifier just begins to distort the signal, then back off the signal strength until distortion is no longer observed. This is the maximum signal the amplifier will handle at maximum volume without distortion. Measure and record the values of the amplifier input and output signals at this level.

3. Measure and record the gain and the maximum input voltage that each stage can handle without distortion.

4. Determine the frequency response of the amplifier by checking the gain of the entire amplifier at each of the following frequencies: 60; 120; 180; 300; 500; 700; and 800 Hz, and 1; 2; 3; 4; 5; 6; 7; 8; 9; 10; 12; 14; and 16 kHz. Graph the frequency response of the amplifier by plotting gain vs frequency.

CHECKING POWER SUPPLY FILTERS

In all ac operated radios, TV, stereo systems, and so on, a process called **rectification** is performed, in which diodes convert the ac of the power line to dc for operating transistors or tubes. The simplest rectification is the **half-wave** depicted in Fig. 6-22. In Fig. 6-22, the rectifier will permit only one alternation, the positive-going one, of each cycle to pass through. The alternations passing through the rectifier are dc, since current flows through the rectifier in one direction only. However, the output is not pure dc. Thus, a filter section made up of electrolytic capacitors and coils or resistors is required to filter out the ac component (**ripple**) of the rectifier output, producing a smoothed dc at the filter output.

One common filter configuration is illustrated in Fig. 6-23. Defects in the filter can be detected by connecting the vertical input terminals of the scope across the filter choke. In many cases, the voltage to be observed will be great enough to permit the signal to be fed directly to the vertical deflection

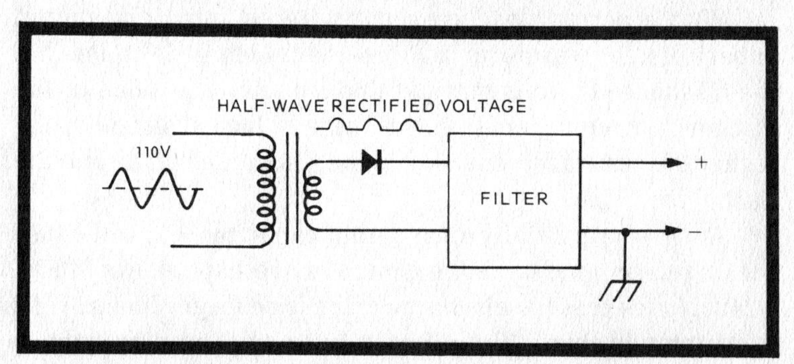

Fig. 6-22. Half-wave rectification produces a pulsating dc.

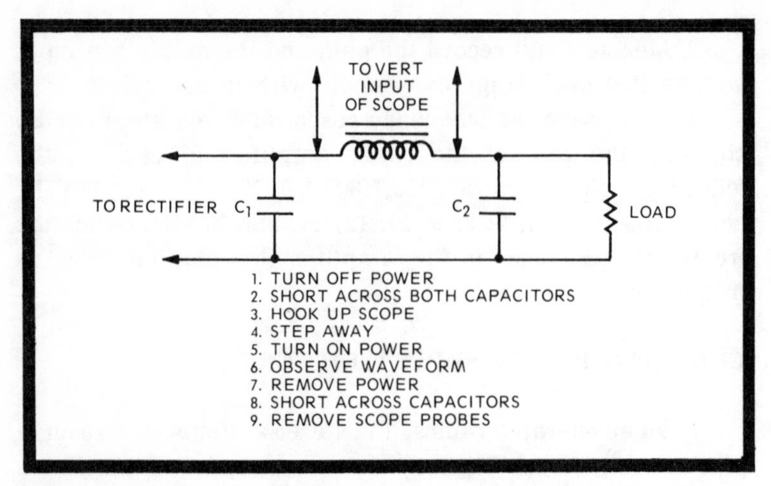

Fig. 6-23. The test setup for checking for filter defects with a scope. Be very cautious when using this setup. Your scope's case might be grounded.

plates. **CAUTION**: many scopes have one terminal of each input grounded to the scope case. The test setup of Fig. 6-23 puts the full power supply voltage on the scope case. Therefore, you should not attempt this test unless you are used to working with high voltages. If you do make the test illustrated in Fig. 6-23, be sure to take every precaution to **avoid contact with any part of the scope**, since this could result in a fatal electrical shock.

A waveform consisting of three alternations should be displayed on the screen for this test. Either the normal power-supply load or a dummy load must be connected across the filter output.

With a normal load current and good input and output capacitors, the waveform in Fig. 6-24A is obtained. Notice that the discharge trace is straight and smooth. The slope of the discharge trace depends on the value of load resistance; the higher the resistance, the slower the discharge will be and the flatter the slope.

An open or partially open output capacitor, C2, will cause the waveform to assume the general waveshape shown in Fig. 6-24B. Notice that the discharge trace is no longer linear. If C2 is completely open, this will usually be evidenced by hum or motorboating in circuits obtaining power from the supply.

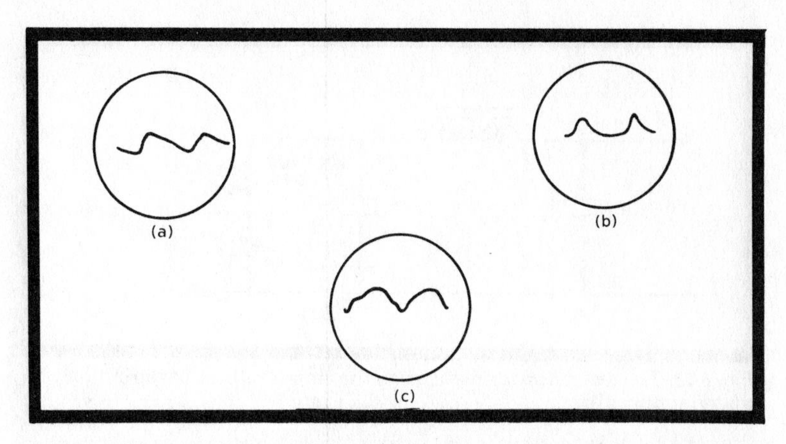

Fig. 6-24. Typical waveforms that might be obtained when checking a power-supply filter.

However, a partially open capacitor that otherwise might go undetected and later cause trouble, can be detected by the scope test. The steepness of the trace in Fig. 6-24A just after the charge peak will depend on the amount of output capacitance still remaining.

A completely open **input** capacitor, C1, will generally be evidenced by low voltage and-or excessive hum. However, this defect can sometimes be anticipated by a scope check of the filter. The pattern in Fig. 6-24C indicates an open or partially open C1. Notice that again the discharge trace is nonlinear, but that now the **charge trace is also nonlinear**. The height of the trace will depend on the amount of input capacitance still left. The worse the condition of the capacitor, the lower the peak of the trace will be.

Insufficient capacitance at either the input or the output of the filter will cause the hum content (**ripple**) of the power supply output to increase above the normal amount. The simplest way to measure the ripple in the output is to compare the ac component of the output with the output of a known voltage source, which serves as a voltage calibrator. If the dc output voltage of the power supply is greater than the input capability of the scope, it will be necessary to connect a voltage divider across the output to obtain a suitable signal for the scope. The test setup for measuring ripple, including the voltage divider, is shown in Fig. 6-25.

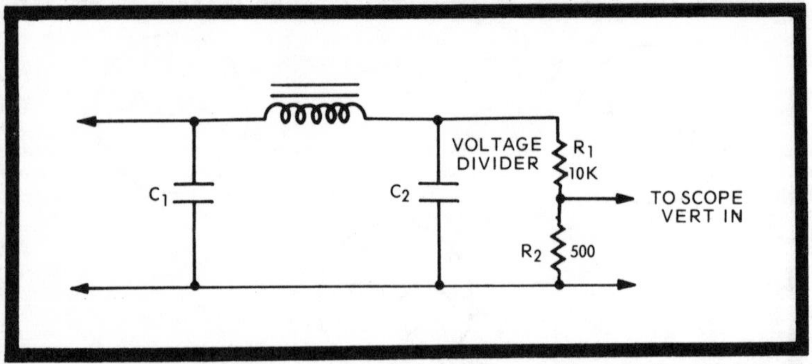

Fig. 6-25. The test setup for measuring the ripple voltage at the output of a power-supply filter.

Connect the calibrator to the vertical input of the scope, using a voltage divider, if necessary. Calibrate the graticule by adjusting the vertical gain so that each minor calibration mark on the vertical axis represents 1 or 2 volts. If the calibrator output is 50V p-p, for example, adjust the vertical gain so that the waveform of this voltage occupies 50 or 25 minor divisions of the vertical axis.

Without changing any of the scope adjustments, disconnect the voltage calibrator and connect the scope's vert in leads to the power supply output, or the voltage divider across the output, if one is used. With the vertical attenuator set to one of its ac positions, the blocking capacitor in the scope input circuit will block the dc component of the power supply output, leaving only the ac ripple. The p-p ripple voltage should then

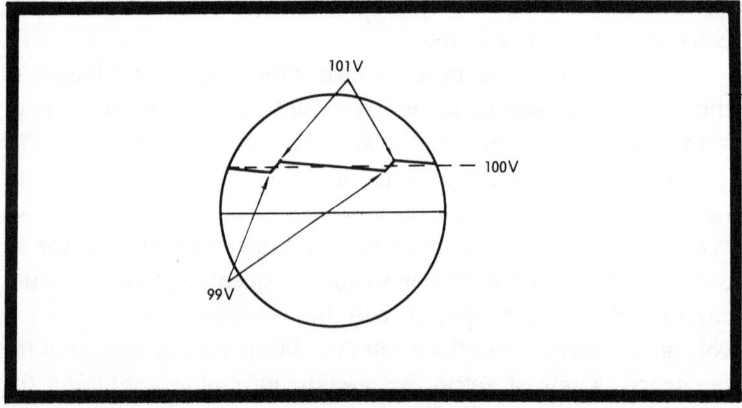

Fig. 6-26. The waveform of a ripple voltage of 2 volts p-p.

be read by noting the number of vertical divisions lines covered by the trace. In Fig. 6-26, the p-p ripple voltage, $E_{p\text{-}p}$, is 2 volts.

Once the dc voltage at the input of the oscilloscope is measured, the **ripple percentage** can be calculated from

$$\text{percent ripple} = \frac{35x\ E_{p\text{-}p}}{E_{dc}}$$

where $E_{p\text{-}p}$ is the ripple voltage measured with the scope and E_{dc} is the dc output voltage of the power supply.

Use the formula and figure the percent ripple for the case illustrated in Fig. 6-26. I got an answer that the ripple voltage is 0.7 percent of the dc output voltage, according to the formula. If that wasn't your answer, try again.

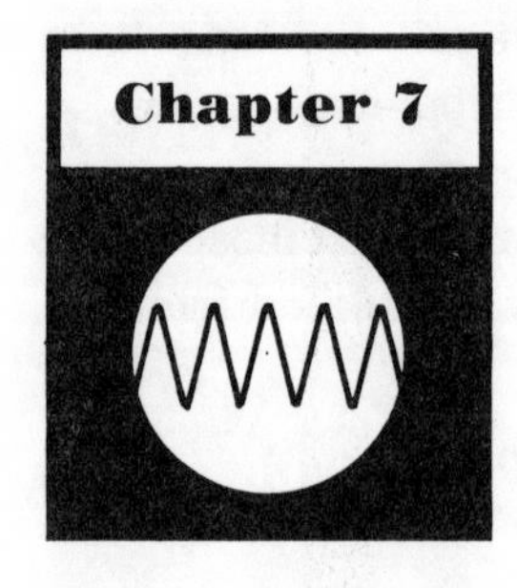

Understanding Scope Specs

In selecting or evaluating a scope, there are a number of characteristics you must consider. The characteristics of any particular scope are given by the manufacturer in a list of specifications, or specs. This chapter will explain the common characteristics of scopes to help you understand the specifications.

AC AND DC COUPLING

Probably the broadest distinction that can be made between various classes of scopes is between ac and dc scopes. Only dc scopes, those scopes having **dc-coupled vertical amplifiers**, are capable of measuring dc voltages and dc components of waveforms.

One measurement that can be made with a dc scope, but not with an ac scope, is the battery measurement illustrated in Fig. 7-1. Notice that the circuit provides a choice of positive, zero, and negative voltages. As the probe of the dc scope is moved downward in the circuit of Fig. 7-1, the trace shifts downward, as shown by Figs. 7-1B, C, and D. This type of setup can be used to calibrate a scope not only for dc measurements, but for peak-to-peak ac measurements as well. Once a good dc scope is calibrated in terms of dc volts, the same calibration holds good for ac measurements.

In displaying a sine wave at the collector of a transistor audio amplifier, we obtain a waveform of the type shown in Fig. 7-2. Since a dc voltage is applied to the collector of the transistor, the sine-wave signal at the collector has a dc component, which is shown as a vertical displacement on a dc scope and in Fig. 7-2. In displaying with a dc scope the video

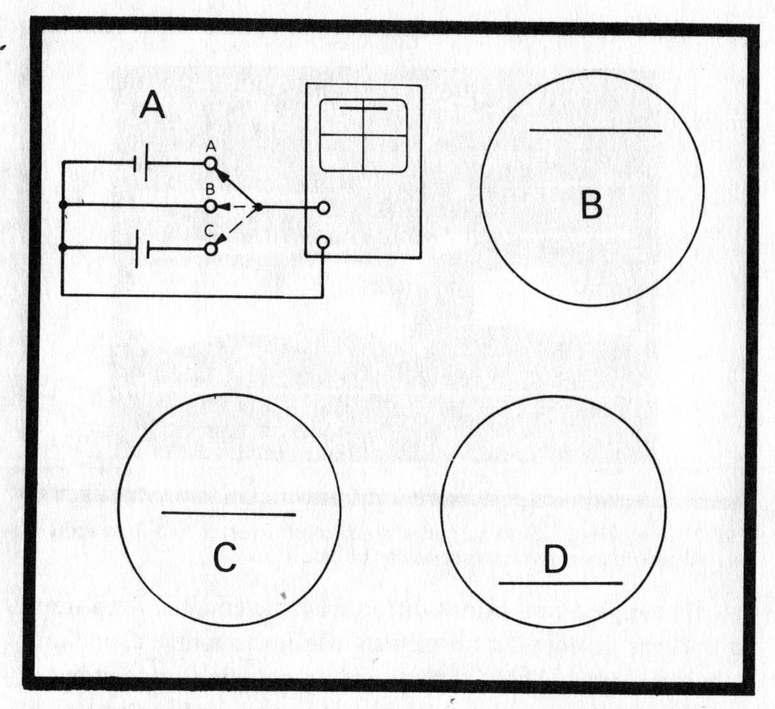

Fig. 7-1. Dc coupled scope is capable of measuring dc voltages.

signal output of the second detector of a television receiver, we see, as in Fig. 7-3, that this signal has a dc component also. Other signals that the dc scope is very useful for displaying are the very low-frequency signals encountered in industrial electronics work, since the dc scope does not attenuate these signals.

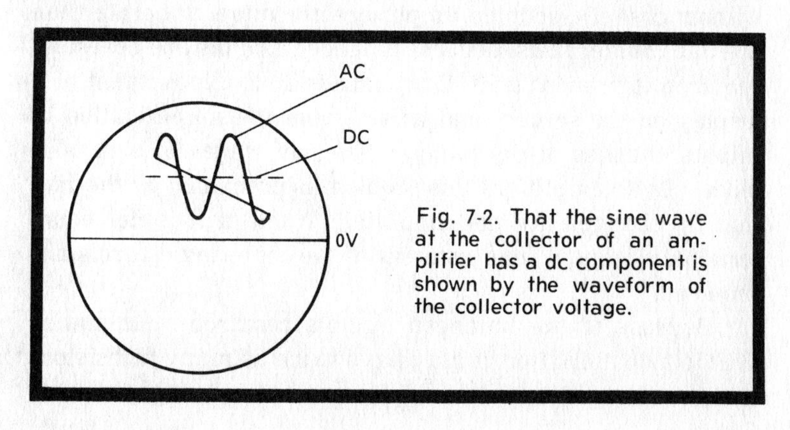

Fig. 7-2. That the sine wave at the collector of an amplifier has a dc component is shown by the waveform of the collector voltage.

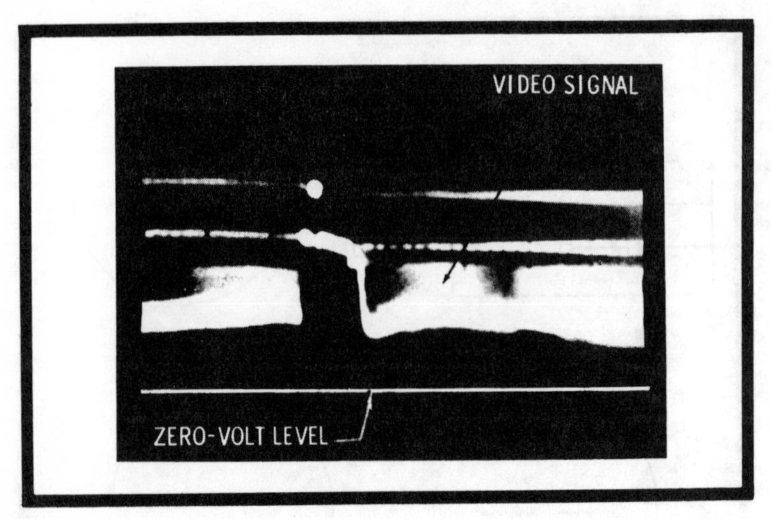

Fig. 7-3. The video signal output of the second detector in a TV receiver has a dc component, which is revealed on a dc scope.

Dc-coupled amplifiers differ from ac-coupled amplifiers in that the dc-coupled amplifiers use no coupling capacitors between stages. That is, they use no coupling capacitor between the collector or plate of one stage and the base or grid of the succeeding stage. Thus, dc voltage changes are transferred from the collector or plate of one stage to the base or grid of the next, and are amplified just as ac signals are. In an ac coupled amplifier, on the other hand, the coupling capacitors used between the stages block dc voltages and prevent the amplification of dc voltage changes.

Direct coupling in a vertical amplifier is not achieved without cost. Dc-coupled amplifiers are more elaborate than their ac counterparts, because balanced circuits are generally required to prevent drift. Drift causes a slow movement of a display on the screen, and arises from the amplification of minute changes of dc voltages in early stages of a dc amplifier. In ac amplifiers this problem is precluded by the fact that dc changes are not amplified. When drift is not counteracted, frequent readjustment of the centering controls are required.

Because of the balanced circuits required to minimize drift in a dc amplifier, it may have twice as many transistors or tubes as an otherwise equivalent ac-coupled amplifier. One bonus of the balanced circuits employed in dc amplifiers is that such amplifiers are less susceptible to instability caused by line-voltage fluctuations. Besides balanced circuitry, dc amplifiers also often use negative feedback to counteract drift. A bonus derived from negative feedback in dc-coupled amplifiers is that it results in improved amplifier linearity and less distortion of signals applied to the amplifier.

In addition to preventing amplification of dc voltages, the coupling capacitors of ac amplifiers adversely affect low-frequency response. That is because capacitive reactance increases as frequency decreases. The effect of coupling capacitors on low-frequency response in an ac coupled amplifier is illustrated in Fig. 7-4. Low-frequency response can be improved by using relatively high values of coupling capacitance, but there is a practical limit to how far this can be carried.

Although some scopes having dc-coupled vertical amplifiers also have dc-coupled horizontal amplifiers, many do not. However, identical vertical and horizontal amplifiers, such as those in the Heath IO-102, are desirable for making phase-shift measurements with a scope to insure that the scope does not itself cause a phase difference between the vertical and horizontal signals. Often, in a scope with identical vertical and horizontal amplifiers, the vertical amplifier will have greater amplification even though it has the same frequency and phase characteristics as the horizontal amplifier, but this causes no problem in phase-shift measurements.

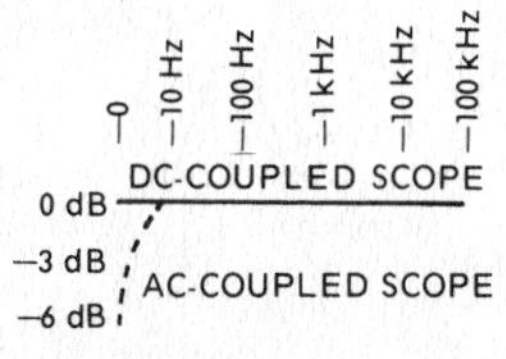

Fig. 7-4. Illustrating how the response of an ac scope drops off at low frequencies due to increasing capacitive reactance of the coupling capacitors.

VERTICAL AMPLIFIER BANDWIDTH

Another broad classification of scopes separates the narrowband and wideband scopes. Scopes considered to have narrowband response generally cover the af range. Those having wideband response usually have vertical amplifiers that are fairly linear over the video frequency range, out to about 5 MHz. The wideband scopes may have dc- or ac-coupled amplifiers. Laboratory scopes, the expensive scopes used in industrial electronics work, may have either narrowband or wideband response, but they very often feature extended high-frequency response, to hundreds of megahertz. Other features of lab-type scopes will be discussed later, since such instruments are becoming more common all the time.

The bandwidth of the vertical amplifier has a great bearing on the usefulness of a scope, especially for viewing nonsinusoidal waves. It also has a great bearing on the cost; usually, the greater the bandwidth, the greater the cost of the scope.

One of the most important changes in the design of scopes in recent years has been an increase in the bandwidth of the

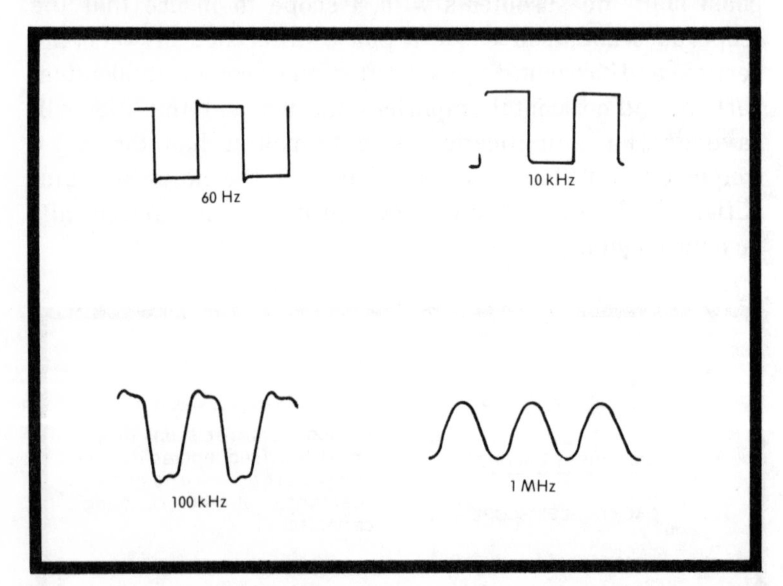

Fig. 7-5. How square waves of various frequencies are presented on a 100 kHz scope.

vertical and horizontal amplifiers. The need for increased bandwidth is based on the types of waveforms that have become standard in recent years. Many modern devices make use of square, triangular, sawtooth, and various pulses and other nonsinusoidal waveforms. The display of these waveforms requires much higher frequency response than the display of sine waves.

An illustration of the ability of scopes with various bandwidths to display complex waveforms is given in Figs. 7-5 through 7-7. Notice in Fig. 7-5 that a scope with a 100 kHz bandwidth is **not** able to give a good representation of a 100 kHz square wave. This is because the harmonics in the square wave are attenuated by the vertical amplifier. They lie far outside the passband of the amplifier. A 1 MHz square wave is reproduced virtually as a sine wave.

Notice that in Fig. 7-6 the 4 MHz scope is seen to give a good reproduction of the 100 kHz square wave, but that it badly distorts the 1 MHz square wave. Only the fundamental and the third harmonic of the 1 MHz square wave lie within the passband of the 4 MHz scope.

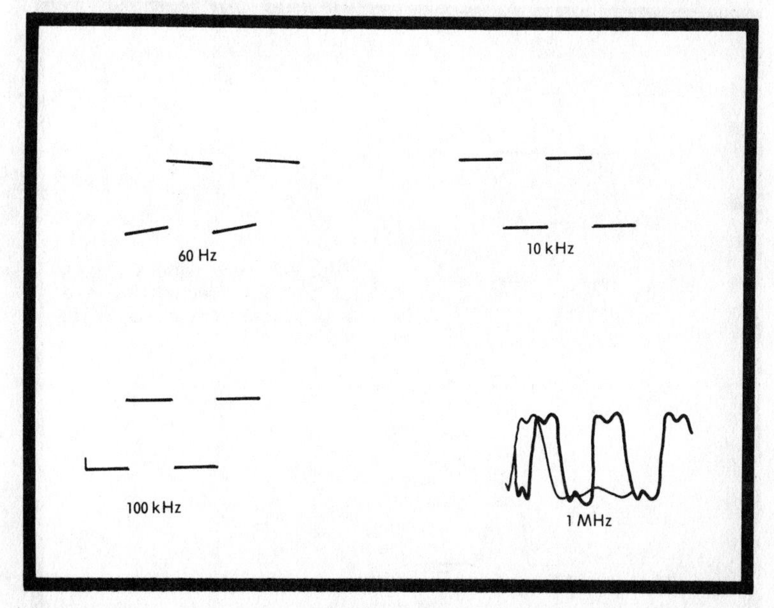

Fig. 7-6. How square waves of various frequencies are presented on a 4 MHz scope.

As can be seen in Fig. 7-7, a 15 MHz lab-type scope is able to give a good reproduction of the 1 MHz square wave. The fundamental of the 100 MHz square wave and the next six odd harmonics are within the passband of the 15 MHz scope. These frequency components are enough to produce a fairly decent square wave.

The frequency response curve of a 5 MHz scope is in Fig. 7-8. This is a curve of the **sine-wave** response of the scope's vertical amplifier. The solid curve shows the response of a dc-coupled amplifier, while the dotted curve shows the response for an ac-coupled amplifier. Notice that the response of a direct-coupled amplifier goes right down to dc.

The frequencies where the output drops to 3 dB below the linear portion of the curve are indicated in Fig. 7-8. At these frequencies, the output sine-wave voltage falls to 70.7 percent of the midfrequency sine-wave voltage. The bandwidth of the amplifier is, by definition, the difference between the upper and lower 3-dB-down frequencies. For a dc-coupled amplifier, the bandwidth is simply the same as the upper 3-dB-down frequency. In Fig. 7-8, this frequency is 5 MHz.

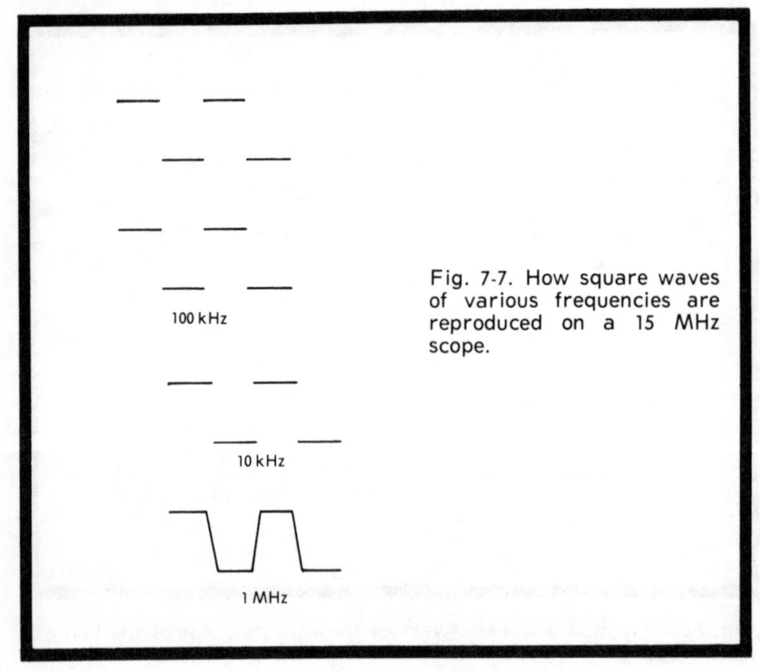

Fig. 7-7. How square waves of various frequencies are reproduced on a 15 MHz scope.

As far as the faithful reproduction of square waves and very brief pulses (transients) is concerned, the high end of the frequency response curve is most important. The shape or "rolloff" of the high end should, ideally, be as shown by the solid line in Fig. 7-8. That is, the high frequency should roll off, or fall off, according to a curve known as a **gaussian** curve. If it does, the 3 dB point will be approximately half the 12 dB frequency. If the upper 3 dB point is 5 MHz, the 12 dB point should be 10 MHz. Lab-grade scopes usually do have a gaussian rolloff. Another thing the response curve should have is a reasonably flat top, within a couple of decibels, to keep some frequencies from being accentuated and distorting the waveform.

In some tube-type scopes, inductances are inserted in the plate circuits of tubes in the vertical amplifier circuits to compensate for interelectrode capacitances and wiring capacitances that tend to shunt high frequencies to ground. This extends the high-frequency response of the vertical amplifier, but causes the response curve to peak at its high end, as shown in Fig. 7-8. For this reason, the inductances inserted for high-frequency compensation are termed **peaking**

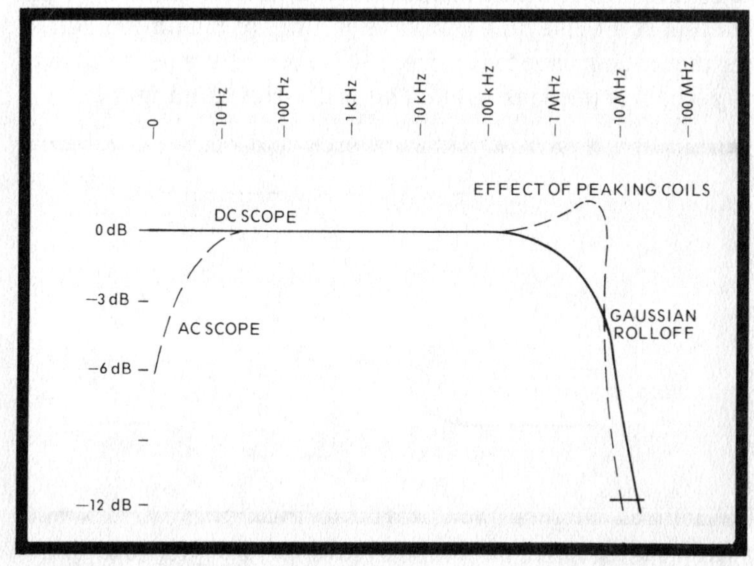

Fig. 7-8. Typical frequency response curves for vertical amplifiers.

coils. The peak in the response curve occurs at the frequency at which the peaking coil resonates with the tube and stray capacitances.

You can use a signal generator having a flat output to check the response of your scope. If you find that the scope has too fast an upper-end rolloff, it probably uses peaking coils.

RISE TIME (TRANSIENT RESPONSE)

For an amplifier to faithfully reproduce pulses with steep leading edges, the amplifier must permit a very rapid rise in voltage. In a vertical amplifier, this allows the crt beam to deflect very rapidly and to follow the nearly vertical edge of a steep pulse. This characteristic of the vertical amplifier is known variously as **rise time, transient response,** and **time response.** It is expressed by a number that is actually an interval of time. Before the rise time, or transient response, of a vertical amplifier can be well understood, it is necessary to understand the steep-sided pulses whose observation depends on this characteristic.

For precision in describing the steepness of the leading edge of a pulse such as a square wave, it is desirable to be able to express the steepness in numbers. This need leads us to the concept of the rise time of a wave, defined as the time required for the leading edge to rise from 10 percent of the peak value to 90 percent of the peak value. This is illustrated in Fig. 7-9.

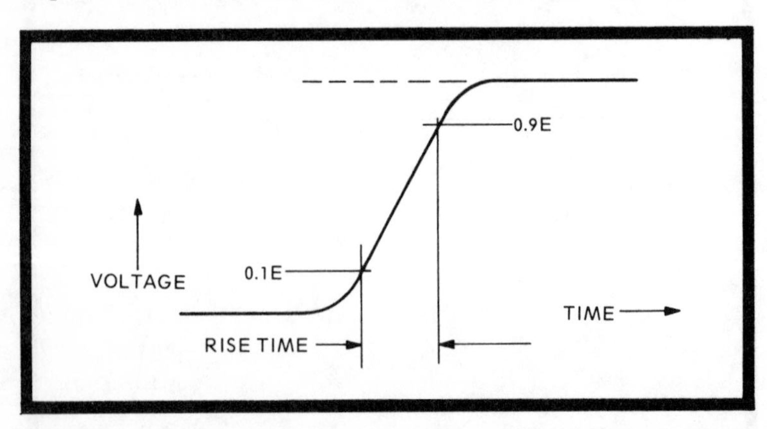

Fig. 7-9. The rise time is the time required for the leading edge of a square wave to rise from 10 percent to 90 percent of its maximum value.

Although the above definition is the generally accepted one, rise time is occasionally taken as the time required for the leading edge to rise from 5 percent of the peak value to 95 percent of the peak value. However, if this or any other definition other than the one given in the preceding paragraph is intended, the intended definition is given along with the rise time.

The rise time of a device that transmits waveforms or displays them is taken as the rise time of the displayed waveform resulting from a theoretically perfect square-wave input. A perfect square wave is impossible to obtain in practice, of course. In practice, the rise time of an amplifier or other device is determined using a square wave whose rise time is even less than the rise time of the amplifier or other device being tested.

One thing of interest for our purposes here is the effect upon rise time when a theoretically perfect square wave is passed through two or more devices in cascade, or series. Suppose that device **a** has a rise time T_{ra}, and device **b** has a rise time, T_{rb}. If a theoretically perfect square wave were fed into the two devices in cascade, the rise time T_{ra} of the output wave should be

$$T_r = \sqrt{T_{ra}^2 + T_{rb}^2}$$

T_r is thus the rise time of the cascade combination of devices **a** and **b**.

For example, if a perfect square wave were fed into an amplifier having a rise time of 3 usec, and if the output of this amplifier were fed into another amplifier having a rise time of 4 usec, the rise time of the output T_r would be about 5 microseconds.

Now, suppose we want to amplify or display a certain waveform, and suppose further that we want the rise time of the output or displayed waveform to be the same as that of the input waveform, within some specific tolerance. Using the chart in Fig. 7-10, which was constructed according to the formula given for T_r, we can find how good an amplifier or scope must be to achieve this result. For example, Fig. 7-10 shows that if we wanted to observe the rise time of a waveform whose rise time is 0.04 usec, we would need a scope whose rise time is not more than 0.01 usec if the error in the rise-time observation is to be kept to less than 3 percent.

Ideally, a scope should have a vertical system capable of rising in about one-fifth the time that the fastest (steepest) signal rises. In such a case, Fig. 7-10 shows that the rise time of the signal (as displayed on the scope) will only be in error by about 2 percent. Vertical systems having a rise time no better than **equal** to the rise time of the fastest signal to be observed are often considered adequate. Whether they are or not depends on the accuracy required. In any case, when the rise time of the scope is known, the rise time of the signal can be calculated from the rise time measured on the screen using the formula

$$T_S \sqrt{T_i^2 - T_o^2}$$

in which T_S is the actual signal rise time, T_i is the indicated rise time, and T_o is the scope rise time. The accuracy with which the rise time can be calculated decreases sharply for

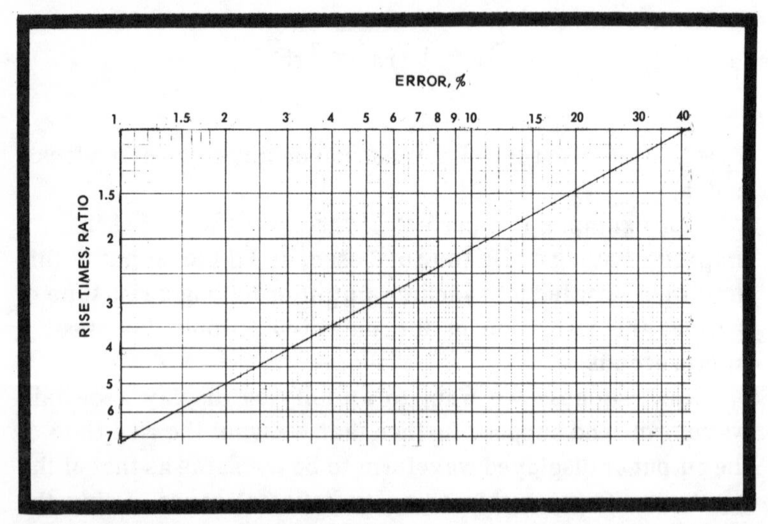

Fig. 7-10. The percent by which the output or displayed rise time exceeds the input signal rise time depends on the ratio of the amplifier or scope rise time to the input signal rise time.

signals that rise faster than the vertical system, because of the increased importance of measurement errors. To illustrate, the following sweep-timing or **display-reading** errors will cause the calculated rise time to be in error by as much as 100 percent.

When $T_o / T_s = 2/1$; 11 percent will cause a 100 percent calculation error.

When $T_o / T_s = 3/1$; 5 percent will cause a 100 percent calculation error.

When $T_o / T_s = 4/1$; 3 percent will cause a 100 percent calculation error.

When $T_o / T_s = 5/1$; 2 percent will cause a 100 percent calculation error.

Thus, if the scope rise time is 5 times the signal rise time, an error of only 2 percent in sweep timing or display reading will result in a 100 percent error in the calculated rise time.

When the fastest sweep is relatively slow compared to the vertical system rise time, the leading edge of the displayed waveform will be steep, and the measurement will be confined to a small portion of the screen. Under these circumstances, the accuracy with which the display can be measured is reduced, and the accuracy with which the rise time of the signal can be calculated is greatly reduced.

Fortunately, very accurate rise time measurements are not required as often as rise time comparisons. For **comparing** the rise time of two signals, a scope rise time no better than the rise time of the signals is usually adequate.

The rise time and high-frequency response of scopes are two of the most important scope characteristics, and they are closely related. Rise time is the more important characteristic for "faster" scopes, and bandwidth is the more important one for "slower" scopes. The product of rise time and frequency should produce a number whose value lies between 0.33 and 0.35, if the scope is to display fast-rising signals without overshoot or ringing. Overshoot, as illustrated in Fig. 7-11, is an excessive initial response to a pulse signal. It is seen in a scope display as a peaking of the leading edge of a pulse. Ringing is a damped oscillation occurring in a signal as a result of an abrupt change in the signal. This defect in a pulse is also illustrated in Fig. 7-11.

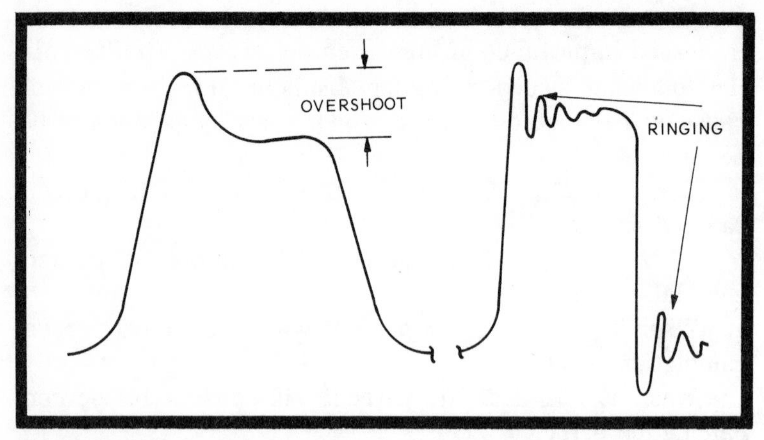

Fig. 7-11. If overshoot and ringing are to be avoided, the product of rise time and an upper 3 dB frequency should be between 0.33 and 0.35.

To illustrate the computation of the product of rise time and frequency response, the product of 0.023 usec rise time and 15 MHz frequency is $0.023 \times 10^{-6} \times 15 \times 10^6 = 0.023 \times 15 \doteq 0.345$. Since for optimum transient response the product should be between 0.33 and 0.35, a scope with the exemplary characteristics will have optimum transient response; that is, it will reproduce steep signal waveforms without significant overshoot or ringing. A factor larger than 0.35 would indicate overshoot greater than 2 percent, and a factor larger than 0.4 would indicate overshoot greater than 5 percent.

You can calculate the rise time of your scope if you know the upper 3-dB-down frequency of the scope, using the equation

$$T_r = \frac{K}{B}$$

where K is 0.35, for an overshoot of less than 3 percent, and B is the upper frequency limit. For example, if the scope response is 10 MHz at the minus 3 dB point, the rise time is 0.035 usec or 35 nsec (35×10^{-9} sec). Such a scope can display a waveform with a rise time of 5 x 35, nsec, or 175 nsec, with a rise-time accuracy of 2 percent. Or, it can display a waveform with a rise time of 3 x 35 nsec, or 105 nsec, with a rise-time accuracy of 5 percent. (See Fig. 7-10).

HORIZONTAL AMPLIFIER CHARACTERISTICS

Horizontal amplifiers are generally similar to vertical amplifiers, and in some cases are nearly identical. However, the performance requirements for horizontal amplifiers are not as stringent as those for vertical amplifiers.

The waveforms generated by the time-base (horizontal sweep) generator make less demand on the horizontal amplifier than signals that may be applied to the vertical amplifier make on it. As a result, the bandwidth of the horizontal amplifier may be about a third the bandwidth of the vertical amplifier.

Besides vertical amplifier response, another factor that limits the maximum presentable frequency of a scope is sweep speed. Since 1 MHz has a period of 1 usec and 10 MHz has a period of 0.1 usec, in order to display a single 10 MHz waveform on the screen, the sweep must take 0.1 usec to span the trace. If it takes longer, more than one waveform will be displayed. A 10 MHz scope having a 1 usec sweep speed means that you will see 10 waveforms on the sweep. The sweep speeds of most triggered-sweep scopes are set by a calibrated control at so many seconds or microseconds per division. The smaller the number, the faster the sweep and the higher the frequency that can be displayed as a single waveform. Many scopes are provided with some form of expansion for even faster sweep speed, making more detailed examination of fast signals possible.

In order to examine one particular part of a waveform, there are two different methods. One is to increase the horizontal gain so that the pattern goes off the screen at both sides, and the trace is thus expanded. Another method, to be discussed further in a later chapter, is to delay the sweep trigger and to use a very fast sweep. This method, **calibrated sweep delay**, has a number of advantages, but it increases the cost of the scope.

The scope is usually used as a high-speed device. There are, however, some applications in which slow sweep speeds are desirable, as in a scope used in surgery to display a patient's heartbeat. This device is called a **physiological monitor**.

The overall speed of response of a scope depends on both sweep speed and on rise time. A figure of merit rating that can be used to express the overall response speed of a scope is the ratio of the vertical system rise time to the time-per-division of the fastest sweep.

Sweep speed is usually continuously adjustable over the total range of the sweep speeds. Usually, continuous coverage between the lowest and highest speeds is provided by a step-type range control, which clicks in at definite speeds, and by a vernier control, which permits continuous coverage between the speeds selectable by the range switch.

The vernier sweep-time-per-division may or may not be calibrated. This control allows us to spread or compress a waveform so that it occupies a certain desired number of graticule divisions, as might be desired for making phase measurements. The control is even more useful if it is calirated, since calibrated sweep allows fractional time measurements to be made without using subdivisions of the graticule scale.

WRITING SPEED

Writing speed, in the strictest sense, is the speed at which the electron beam moves across the face of a crt in tracing out a waveform. The writing speeds that are important scope characteristics are **photographic writing speed** and **stored writing speed**. Photographic writing speed is the maximum speed at which the spot of light can move across the face of the crt and still be photographed. Stored writing speed is the maximum speed at which the spot of light can move across the face of the crt and still be presented on the screen of a storage scope. That is, it refers to the linear velocity of the trace over the face of the viewing screen, and this speed is typically expressed in inches or centimeter per second.

A storage scope, more about which later, holds a display on the screen for an indefinite length of time until the operator removes the trace by pushing an erase button on the front panel. Writing speed is important to us mainly when considering screen photography or storage scopes.

When photographic writing rate is stated, there should also be included a description of the photographic equipment used. At least the camera and film type should be specified. The f number and magnification of the lens may also be stated.

High-quality scopes such as the 7000-series by Tektronix, Inc., have typical writing speeds of 5000 cm per usec. At this rate, a scope could draw a line around the equator in 1 second! Stored writing rates are less than this because a storage crt takes longer to pick up, or freeze, an image than photographic film takes. The fastest stored writing speed yet achieved is the 200 cm per usec of the Tektronix type 7623.

The writing speed determines the highest frequency waveform or the fastest pulse that can be photographed from the screen or stored on a storage crt. In order to find out whether a given signal can be photographed or stored using a particular scope, it is necessary to determine how fast the writing beam of the scope must travel to trace out the waveform of the signal, and then to compare this speed to the photographic or stored writing speed of the scope. If the photographic or stored writing speed is greater than the required writing beam speed, the signal can be photographed or stored.

For sine waves, the beam speed is related directly to the frequency and amplitude of the applied signal. The speed is not constant, but varies, and the greatest speed occurs when the sine wave passes through its zero-voltage points, or baseline. The equation for the maximum writing velocity (V_m) of a sine wave is

$$V_m = 2\pi f A$$

where A is the amplitude of the wave in centimeters or inches.

If a 1 MHz sine wave has an amplitude of 2 cm p-p, A equals 1 cm, and the maximum writing speed in cm per sec will be 6.28 times 1 times 10^6, or 6.28 cm per usec. In order to photograph or store this signal, a scope having a photographic writing speed or a stored writing speed in excess of 6.28 cm per usec would be required.

For step signals—square waves, pulses, etc.—the required photographic or stored writing rate depends on the sweep speed required to display the signal. In a triggered-sweep scope, the required writing speed is simply the reciprocal of the time-per-division setting. If the time-per-division setting is 2 usec per division, the horizontal sweep speed is 1 division divided by 2 usec, which is 0.5 div per usec, where divisions are in centimeters or inches. If the time-per-division setting were 2 usec per div, then a scope would have to have a photographic or stored writing speed greater than 0.5 division per microsecond in order for the signal to be photographed or stored.

The speed at which the writing beam of a recurrent-sweep scope must travel to display a rectangular pulse can be determined by dividing the width of the pulse in divisions by the time duration of the pulse. The speed calculated can be converted to centimeters per microsecond, or whatever units are needed to compare it to the photographic writing speed specification of the scope. We do not mention stored writing speed here because storage scopes use a triggered sweep.

VERTICAL SENSITIVITY

This characteristic is related to the deflection sensitivity of the crt and to the gain of the vertical amplifier. The deflection sensitivity is the amount of deflection caused by 1V applied to the deflection plates. This is an inherent characteristic of the crt, and for most tubes is about 0.1 cm per volt. In other words, 1V on the deflection plates produces 0.1 cm of crt beam deflection. The deflection sensitivity of the crt limits the minimum and maximum signals that can be measured. Also, the deflection sensitivity limits practical direct measurements to 1V minimum signal, since smaller signals would cause a display amplitude even less than 0.1 cm. Deflection sensitivity and screen size also fix the maximum measurable level. If the voltage at the deflection plates times the deflection sensitivity is greater than the screen size permits, the beam will be deflected off screen.

Inserting a probe and an amplifier between the signal source and the vertical deflection plates increases deflection sensitivity and isolates the signal source and the crt. This vertical deflection system, by including an amplifier, implies **voltage gain**. Vertical amplifiers increase deflection sensitivity. They increase the deflection that a signal voltage can cause. Also, the amplifiers contain calibrated step control of gain and attenuation, thereby extending input deflection sensitivity from, perhaps, 0.04 div/v to 100 div/v in several steps.

Since division-per-volt terminology usually requires calculation for practical observations, vertical deflection sensitivity is usually expressed in terms of **deflection factor**. Deflection factor is the inverse of deflection sensitivity, and is expressed in volts per division. Here's an example: If a crt has a deflection sensitivity of 0.1 cm per volt, it has a deflection factor of 10 volts per cm. If the amplifier has a gain of 1000, the vertical deflection sensitivity will be increased to 100 cm per volt, and the deflection factor, the sensitivity of the scope itself, will be 0.01 volt per cm, or 10 mV per cm.

If the specs for a scope give the vertical sensitivity as 10 mV per cm, this means that an applied signal of 10 mV at the vertical input terminals will produce a 1 cm vertical deflection with the vertical gain turned all the way up.

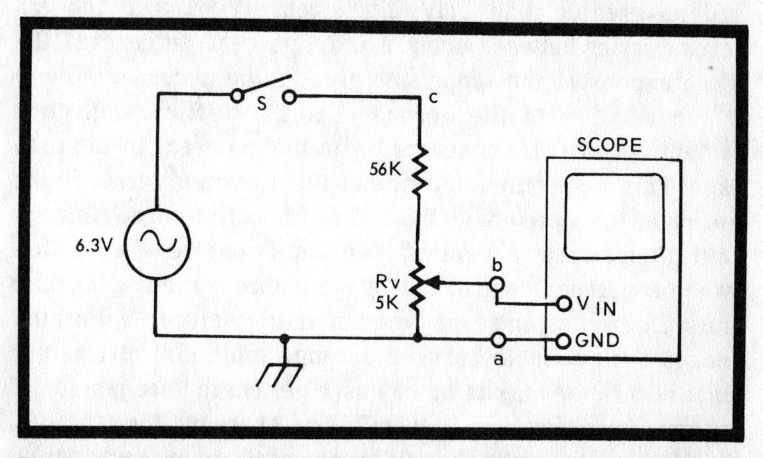

Fig. 7-12. Circuit for use as a voltage source for determining scope sensitivity.

If you do not know the vertical sensitivity of your scope, you can determine it by use of the following procedure.

1. Connect the circuit of Fig. 7-12, connecting the output of the voltage divider to the vertical input of the scope to be checked.

2. Set the vertical attenuator and vertical gain controls for maximum vertical gain. Adjust voltage divider level control R for a signal giving 1 cm of vertical deflection on the crt.

3. Open switch S and measure the resistances Rab and Rbc. Substitute the measured resistances in the formula

$$\text{mV per cm, sensitivity} = \frac{R_{ab} \times 6.3 \times 2.828 \times 103}{R_{bc}}$$

ACCURACY

Vertical accuracy is a measure of the distortion produced in the vertical deflection system, and is stated as a percentage. If a scope has a vertical accuracy of 3 percent—a very good accuracy, by the way—then once the scope is calibrated, it should be possible to measure voltages with an accuracy of 3 percent or better. If, for example, a scope with a vertical accuracy of 3 percent is calibrated for 20V per div, a deflection of 5 div will indicate an amplitude no less than 97V and no greater than 103V. The accuracy of good modern scopes varies between about 2 and 5 percent. In general, the more expensive the scope, the greater the accuracy.

In addition to the accuracy of the vertical deflection system, we are also concerned with the accuracy of the time base. This is specified in terms of the maximum error in the timing of the sweep over the full width of the sweep, for any calibrated sweep. A calibrated sweep is one selected with a sweep frequency control that gives a direct reading in time units. Depending on scope type and manufacturer, calibrated ranges are available between a range minimum of 1 ns per cm, and a range maximum of 1 usec per cm to 5 sec per cm. If the time-base accuracy is specified as 3 percent, for example, the time required to make a full horizontal sweep should not be less than 97 percent nor more than 103 percent of the time indicated by the sweep-speed setting. If, then, the sweep speed is set at 1 usec per cm, and if the full-scale sweep is 10 cm, a 3 percent time-base accuracy means that the sweep may cover the full-scale distance in as little as 9.7 usec, or as much as 10.3 usec. Time-base accuracy is important when accurate time measurements are to be made.

It is also important to make a distinction between time-base accuracy and **sweep linearity**, to which time-base accuracy is related. Time-base accuracy is highly dependent on sweep linearity. However, basing the accuracy on the full-scale sweep has the effect of averaging the various rates of horizontal sweep that may occur during one complete sweep. There are various kinds of sweep nonlinearity, and several of them may be present in the same scope. The most common involves slowness at the beginning and end of the sweep, with the fastest sweep rate occurring near the center of the sweep. Since there is no generally accepted way of specifying linearity, it is not included in the specifications of a scope. Linearity may be checked by any of several methods explained in the sweep linearity section of Chapter 5.

INPUT IMPEDANCE

The input impedance to the vertical amplifier can be simulated by a high resistance shunted by a small capacitance, as in Fig. 7-13. Since capacitive reactance varies with frequency, in order for the impedance spec to hold for all frequencies, it is expressed in terms of resistance and

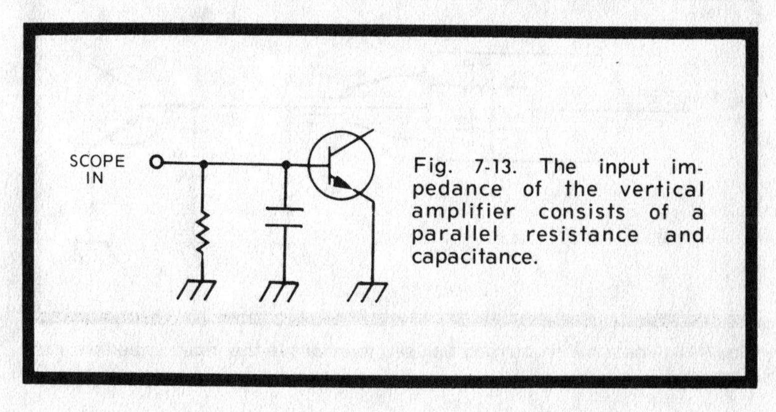

Fig. 7-13. The input impedance of the vertical amplifier consists of a parallel resistance and capacitance.

capacitance, rather than in terms of resistance and capacitive reactance. It may be referred to by such names as input RC, and input time constant. The resistive part of the spec is typically 1 megohm, and the capacitive part is typically 15 to 50 pF.

In some applications, even this high resistance and small capacitance may produce undesirable loading of the circuit whose waveforms are being observed. In other words, the loading can cause different waveforms to be displayed than would exist with the scope disconnected, and thus can give a misleading presentation. To minimize this loading, a passive probe, that is, one containing no amplifying device, may be used with the scope. Such a passive probe may consist of a parallel resistor and capacitor, as in Fig. 7-14. The result of using the probe is that there is connected to the circuit being investigated a new effective loading capacitance smaller than the original capacitance and a new effective loading resistance larger than the original resistance. Thus the loading effect of the vertical input circuit is reduced through the use of a probe, and the input resistance might be increased to 10 megohms, say, and the capacitance might be decreased to perhaps 10 pF.

CRT SIZE AND VOLTAGE

Two important factors affecting the viewability of a scope are crt size and voltage. First let's consider the importance of crt size.

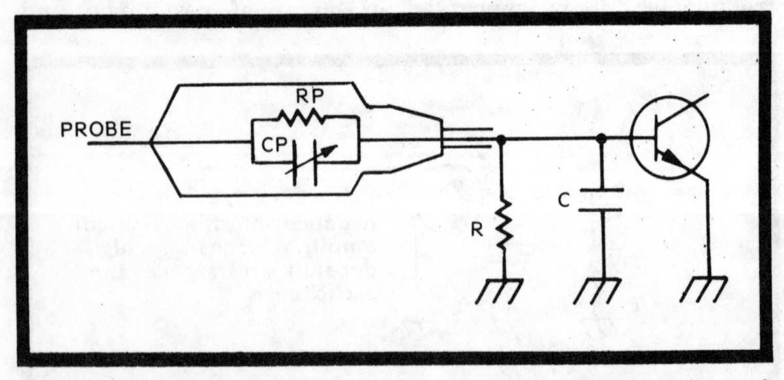

Fig. 7-14. A passive probe may be used to improve the input impedance of a scope.

The diameter of scope screens varies from 3 to 7 inches, with 5 inches being a very popular size. Sometimes, as in the Heath IO-102, a circular screen will be masked to give a rectangular viewing area. There are also rectangular scope screens, and they seem to be rapidly gaining in favor, especially in advanced lab-type scopes. Five inches is also a common size for rectangular screens.

One important fact to note is that screen size along with deflection sensitivity fixes the maximum measurable voltage level. For example, if a crt has a deflection sensitivity of 0.1 cm per volt, and a signal of 60V p-p is applied to its vertical deflection plates, the amplitude of the resultant display will be 6 cm. Assuming a rectangular viewing area, the height of the viewing area will have to be greater than 6 cm to view the signal.

Because of the curvature of the face of a crt, especially at its edges, the entire screen is not useful for display purposes. In fact, the maximum useful screen diameter is only about 80 to 90 percent of the rated screen diameter. Since different tubes have different degrees of curvature, only an approximation of the useful screen diameter can be given.

A crt is sometimes classified according to the voltage applied to the second (accelerating) anode. In general, a crt operated with second-anode potentials of less than 2000V is considered a low-voltage tube, while those operated with higher potentials are said to be high-voltage tubes.

In some ways high-voltage operation is more advantageous than low-voltage operation, and in some ways the reverse is true. The low-voltage tube has an inherently high sensitivity because of its low-velocity beam. On the debit side, the low-velocity beam produces a trace of relatively low brightness, a trace that may not be readily visible in a brightly lighted room, especially if the ambient light falls directly on the screen.

The relatively low brightness of displays on low-voltage tubes does not mean that these tubes lack utility. On the contrary, the average scope using a low-voltage tube will produce a trace which is quite legible under semishaded conditions. Also, the films available nowadays permit photographing this trace with little difficulty. Finally, the advantage of high sensitivity often outweighs high trace intensity in everyday applications.

However, some scope applications demand viewing in broad daylight. In these applications, high trace intensity is very desirable. Also, high writing speeds demand a bright trace, because the faster the beam moves across the screen, the less time it spends at any given point on the screen, and the less energy it can impart to that point to make it glow. Thus, high writing speed demands high beam velocity, which in turn means that a high accelerating voltage is required.

Furthermore, the higher the velocity of the beam electrons, the less the possibility of the beam being deflected by stray fields. Another advantage of the high-velocity beam is that it is narrower and produces a smaller spot on the screen. Thus, a high-velocity beam results in greater resolution, or more detail, in the display.

OTHER FACTORS

In addition to the considerations already mentioned, there are a number of other important factors to be considered when evaluating a scope.

One important factor is circuit design. A solid-state design affords extra reliability, insuring long periods of operation without adjustments. By using worst-case circuit design, the number of calibration adjustments can be held to a minimum. Zener diode protection or some other method of protecting against excessive input voltages is also a very desirable feature. Besides being protected, the input should offer a high input impedance to minimize loading of the circuit under test. FET devices in the input stage offer such a high input impedance.

Power consumption is also important, and not just from the expense of operation. Low power consumption makes vent holes in the case unnecessary. Eliminating vent holes protects inner circuits from dirt and other contaminants that might be encountered in the field. Low power consumption and the concomitant cooler operation mean that greater environmental temperatures can be tolerated.

In addition to the performance advantages it offers, solid-state design makes for smaller size, lighter weight, and greater portability.

The cost of scopes varies widely, ranging all the way from about $100 for a service-type scope to over $4000 for some lab-type scopes. The cost is closely tied to the bandwidth and accuracy. It also depends on the operating features of the scope. Such features of advanced scopes as delayed sweep facilities, multiple-trace capability, sampling capabilities, storage provisions, and differential input, add greatly to the cost. Since these features are expensive, it is important to understand them and know which ones are needed for your scope application. In the next chapter we will explain some of the features of advanced scopes.

Advanced Scopes

There are a great many tests and measurements that can be made with a simple service-type scope, and we have already discussed some of them. However, there are many other tests that either require that the scope have special capabilities, or are facilitated or enhanced by such capabilities. The special capabilities discussed in this chapter, or most of them, are available to the engineering technician in lab and in industrial applications, and are becoming increasingly available to the service technician. Not every advanced scope will have all of the features explained in this chapter, but most will have some.

DUAL-TRACE AND DUAL-BEAM SCOPES

A very useful ability for a vertical amplifier to have is the ability to pass either one of two input signals to permit viewing either signal, without disturbing connections, so the two signals may be compared. Manual switching is possible and is the simplest method of switching between two signals, but electronic switching permits viewing the two signals at the same time. Since each signal traces out a separate display, scopes with built-in electronic switches are often called **dual-trace** scopes. These should not be confused with **dual-beam** scopes. In a dual-trace scope, a single beam in a crt is shared by two signal channels. In a dual-beam scope, however, the crt has two electron guns, which produce two beams, one for each signal channel.

For some purposes dual-beam scopes are superior to dual-trace scopes, and for other purposes, vice versa. A dual-beam scope may display two simultaneous, nonrecurrent signals of short duration, but a dual-trace scope may not. Furthermore, some dual-beam scopes can display nonrecurrent signals on **different time bases**. The main advantages of dual-trace scopes are lower cost and inherently better waveform comparison capability.

Steady display of two signals that are nonsynchronous with each other is possible with a dual-trace scope. This is possible because the triggering signals may be switched along with the input signals. Such a display might be used in a situation where one waveform is some kind of standard. Dual-beam scopes also permit this type of comparison, provided they have two sweep generators and two sets of horizontal deflection plates.

There are two ways in which electronic signal switching is accomplished in scopes: the **chopped mode** and the **alternate mode**. The chopped mode was employed in the electronic switch accessory discussed earlier. This dual-trace mode is most useful at low sweep speeds and when it is desired to have dual-trace display when using 60 Hz or external sweep provisions.

Dual Chopped Mode

The **dual chopped mode** uses a square wave of 100 kHz or more to switch the vertical input between the two input channels. The two input channels are designated channels A and B, or channels 1 and 2; and their signals are displayed one above the other.

Dual Alternate Mode

In the dual alternate mode of operation, the signals are also displayed one above the other on alternate sweeps. On one sweep, the signal on channel A is displayed. Because of the frequency at which the switching takes place, phosphor persistence, and persistence of vision, the two signals appear to be displayed continuously.

Alternate Mode

The alternate mode is used more frequently and is preferred for displays employing fast sweeps. The chopped mode is usually reserved for comparing low-frequency recurrent signals, or for comparing long-duration nonrecurrent signals.

Sometimes, when displaying two bright traces using the chopped mode, the display may show the chopping waveform transients as faint lines connecting the two traces. In some scopes, the crt is turned off (blanked) during the transition intervals to prevent the transients from affecting the display.

The chopping frequency should be as high as possible so long as the resulting traces are not broadened significantly by distortion of the chopping signal. When the chopped mode is used with relatively fast nonrecurrent sweeps, the traces are not continuous, but are made up of a number of segments, the number depending on the chopping frequency and the sweep duration. For example, if the chopping rate is 1 MHz and the sweep duration is 0.1 msec, there will be 100 segments in each trace. How well these segments depict the input waveforms determines the limits of usefulness of the chopped mode compared to the alternate mode.

One of the advantages of dual-trace (or dual-beam) display is illustrated in Fig. 8-1. If the voltage waveform from a circuit is applied to channel A and the current waveform is applied to channel B, the phase difference is shown directly on

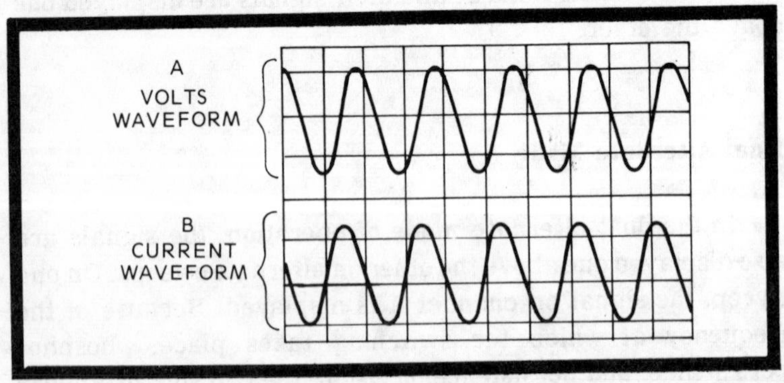

Fig. 8-1. The phase difference between voltage and current may be shown directly on the screen of a dual-trace scope.

the screen. Direct measurement of the phase difference between the input and output of an amplifier is also possible with a dual-trace display. If the input signal to the amplifier is applied to channel A and the output signal of the amplifier is applied to channel B, the input and output waveforms are displayed one above the other on the screen, and a phase comparison is readily made.

The main advantage of a triggered-sweep scope as compared to a recurrent-sweep scope is its capability of beginning the sweep at exactly the same point on each trace. This allows very accurate comparisons to be made between two signals of the same frequency, provided that the signals can be displayed simultaneously. Unfortunately, many triggered-sweep scopes do not permit this comparison, because they are only single-trace instruments. In order to

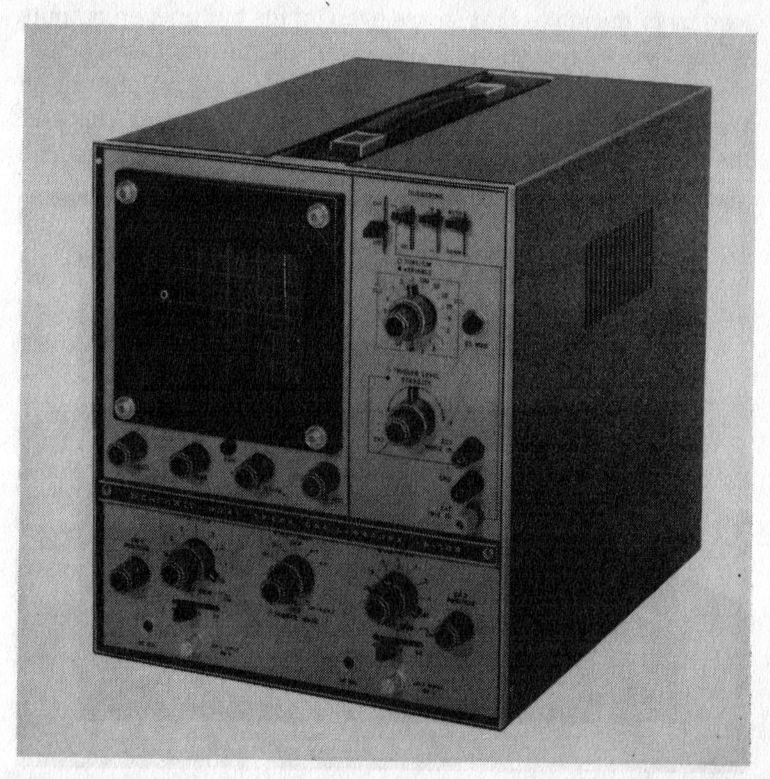

Fig. 8-2. A modern dual-trace triggered-sweep scope, the Heath IO-105. (Courtesy Heath Co.)

realize the full potential of the triggered sweep, it must be combined with dual-trace capability. In short, dual trace and triggered sweep go together. A good example of a dual-trace triggered-sweep scope is the Heath IO-105 pictured in Fig. 8-2.

Dual trace is very useful for troubleshooting purposes. Suppose you have restored the operation of a stereo amplifier, but find that the balance control must be turned off center toward the left channel to get equal output from both channels. A single trace would permit observation of each waveform, but a comparison of the signals on a dual-trace scope would facilitate troubleshooting. A check at the output would disclose that the left-channel signal, shown on the lower trace in Fig. 8-3, has a much lower amplitude. Working back through the stereo amplifiers, with one channel connected to the scope A input and the other channel connected to the B input, would lead us to the stage that shows even inputs but uneven outputs of the two stereo channels. Looking at the emitter of both stages, as shown in Fig. 8-4, we would see a signal present on the left channel but not on the right channel. A quick check of the emitter bypass capacitor might show that it is open. The

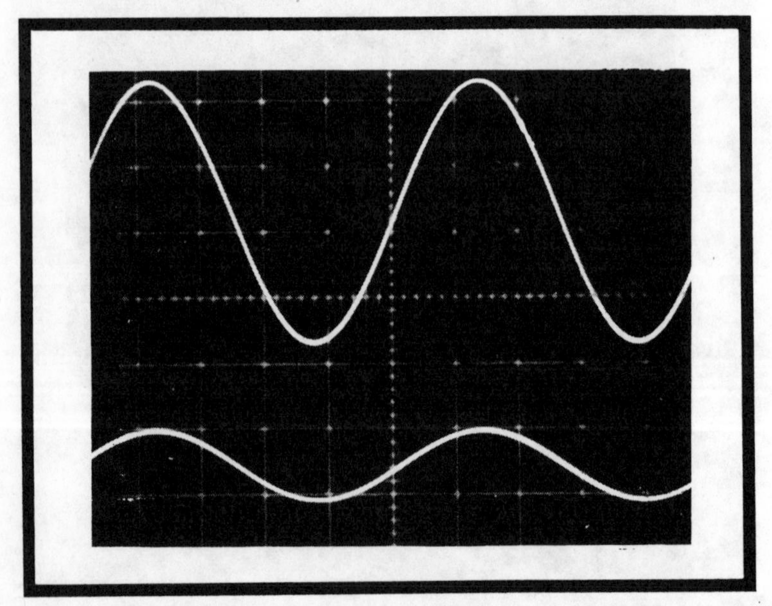

Fig. 8-3. A dual-trace display of stereo signals in which the left channel (lower trace) signal is shown to have a much lower amplitude.

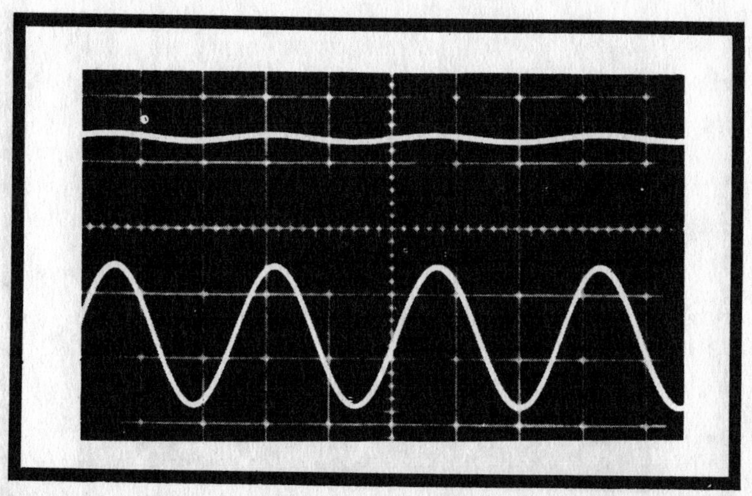

Fig. 8-4. Unequal emitter signals disclose the source of trouble in a stereo amplifier.

signal observed across the emitter resistor in the left channel represents degeneration, a source of reduced gain. With the emitter resistor properly bypassed, almost no signal should be observed across it.

As another example of the use of a dual-trace scope for servicing, suppose the flesh colors are wrong on a color television receiver and you know that both demodulation signals are present. It might be that the demodulation signals do not have the proper phase relationship. They should be 90 deg out of phase. Connecting a dual-trace scope to the demodulation grids might result in a display such as that in Fig. 8-5, indicating that the phase of the 3.58 MHz demodulating signals is much less than 90 deg. If this were caused by, say, a leaky capacitor in the 3.58 MHz phase shift network, replacing the capacitor would result in the display in Fig. 8-6. This pattern shows the proper relationship of the demodulating signals.

Fig. 8-7 is a closeup of the vertical mode selector switch of the Heath IO-105. When the switch is in the CH 1 position, only channel 1 signals are displayed. Similarly, when the switch is in the CH 2 position, only channel 2 signals are displayed. In the CHOP position, channel 1 and channel 2 signals are displayed alternately at 10 usec intervals. This mode is not

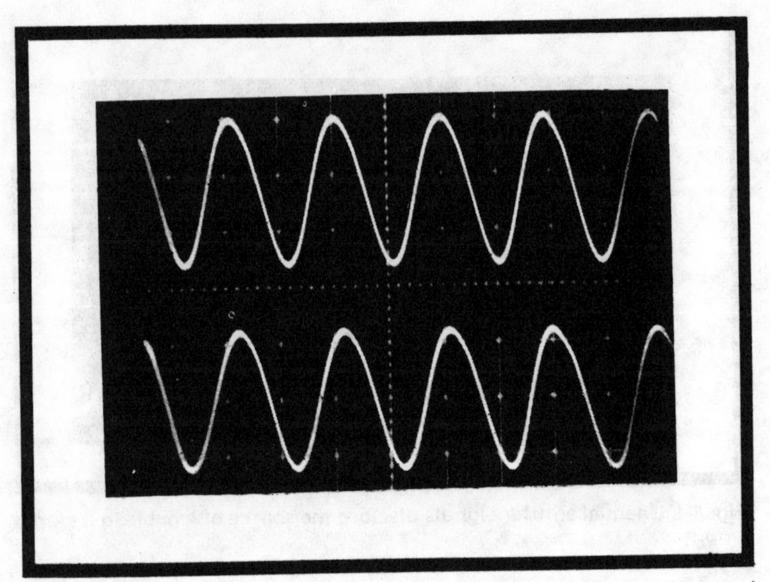

Fig. 8-5. A dual-trace comparison of color TV signals shows incorrect phase angle.

suitable for sweeps faster than 1 msec per cm. At slow sweep rates, switching occurs so many times during the sweep that the waveforms in the display have no perceptible gaps. However, at fast sweep rates, the gaps in the waveforms become visible, as shown in Fig. 8-8, in a real-time illustration of a chopped dual-trace display when the sweep rate is too fast. As a rule, the chopped mode is not suitable for sweeps faster than about 1 msec per cm for this reason.

When the vertical mode switch is in the ALT position, channels 1 and 2 are displayed on alternate horizontal sweeps, as illustrated in Fig. 8-8. In this mode of operation, switching occurs during retrace. The persistence of the crt phosphor and the persistence of vision cause each waveform to appear even while the other is being traced out, although this isn't apparent from Fig. 8-8. If the sweep time is sufficiently long, the waveforms will flicker, since there is a limit to the persistence of the crt and of vision. If the sweep time is longer still, the alternate displays will be clearly seen, as in Fig. 8-8.

When the vertical mode switch of Fig. 8-7 is in the X-Y position, the signal on channel 1 is displayed on the vertical (Y) axis, and the signal on channel 2 is displayed on the

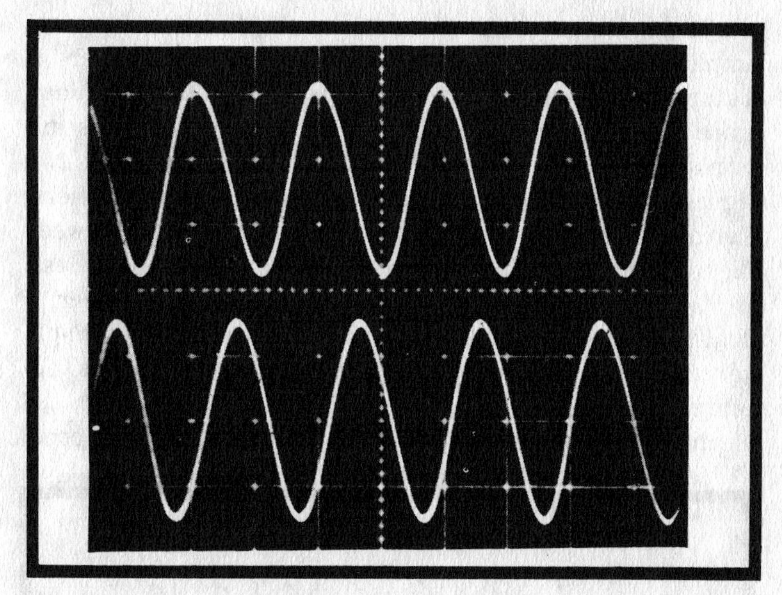

Fig. 8-6. A dual-trace display of color demod signals showing the proper phase relationship between the signals.

horizontal (X) axis. This mode is useful for producing Lissajous figures, for phase measurements, and for other applications where a voltage is to be presented with respect to another voltage instead of with respect to time. As previously explained, X-Y displays are also possible with an ordinary single-trace scope. In displaying Lissajous figures, you will recall, one of two frequencies was applied to the vertical input, and the other was applied to the horizontal input. Thus, one signal was amplified by the vertical amplifier and one was amplified by the horizontal amplifier. Ideally, if two signals

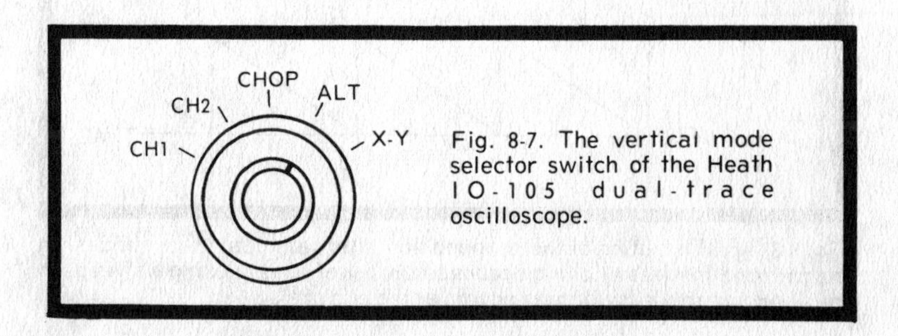

Fig. 8-7. The vertical mode selector switch of the Heath IO-105 dual-trace oscilloscope.

are to be compared, they should be amplified by identical amplifiers. However, the horizontal and vertical amplifiers of a scope are not identical. The two vertical amplifiers of a dual-trace scope are identical. Thus, when Lissajous figures are displayed on the X-Y mode of a dual-trace amplifier, the two signals to be compared are amplified by identical amplifiers. Furthermore, there is only a very small phase shift between the matched X and Y amplifiers, typically 3 percent or less.

Why two channels for dual trace? As you saw earlier, a dual-trace display can be produced on a scope having a single vertical amplifier, using an external switch. This is accomplished by using a square-wave signal, developed in the electronic switch, to connect the input signals to the scope.

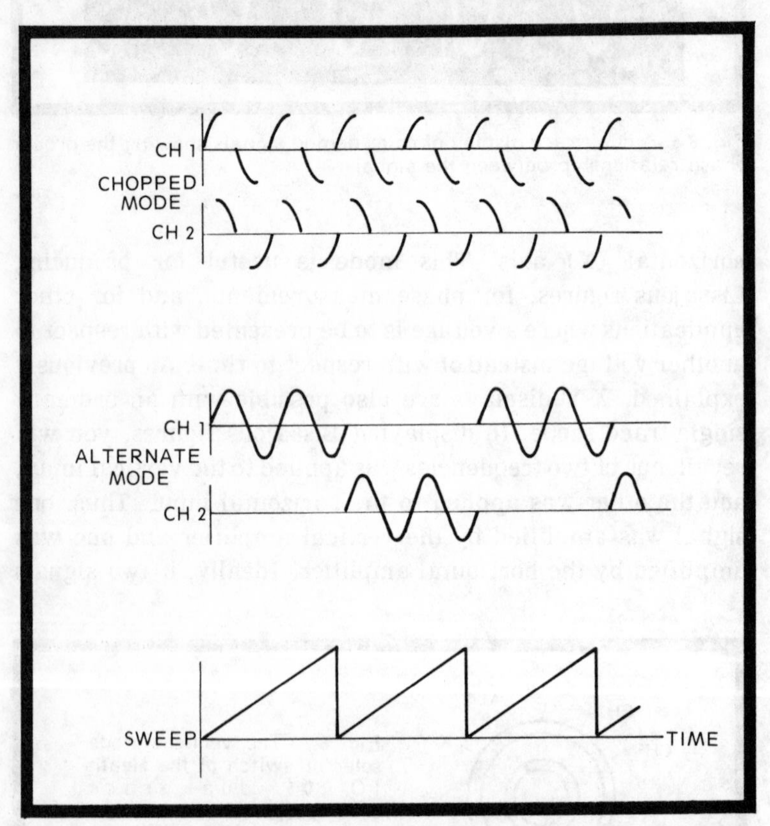

Fig. 8-8. An illustration of the chopped and alternate dual-trace modes. In the chopped mode, switching occurs a number of times **during** a sweep. In the alternate mode, **switching** occurs at the end of a sweep.

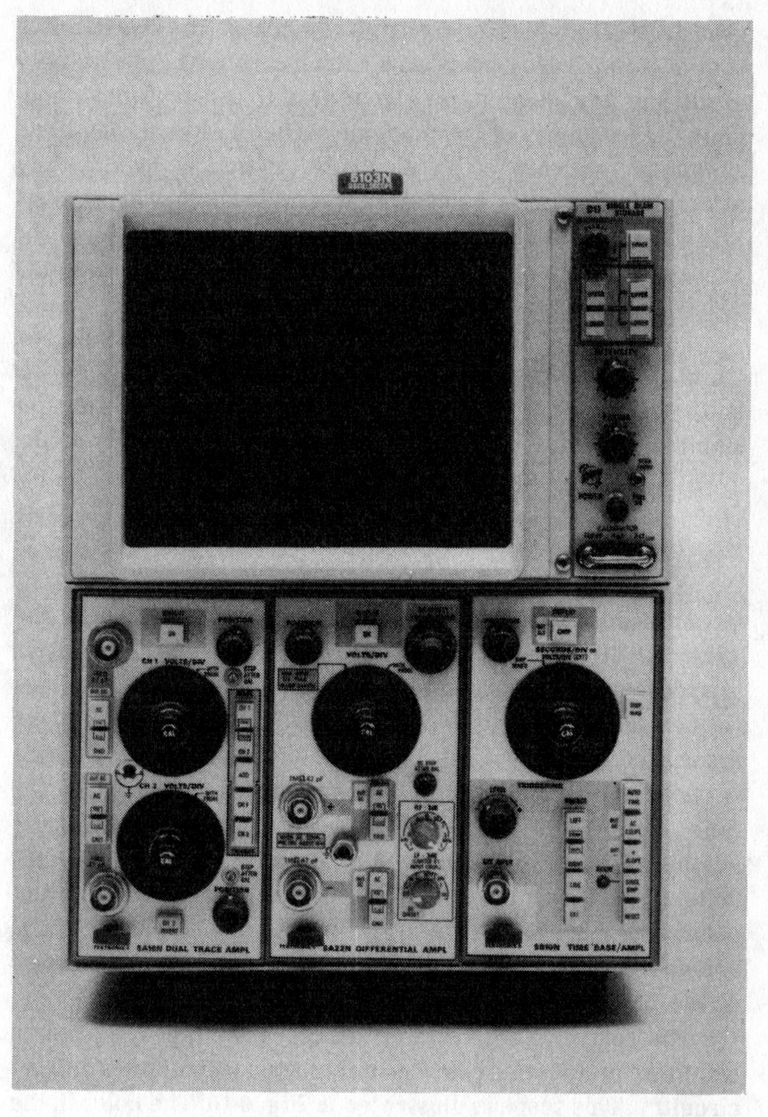

Fig. 8-9. The Tektronix 5103N plug-in scope consists of a mainframe, one or two vertical preamplifier plug-ins, and a sweep generator plug-in. (Courtesy Tektronix, Inc.)

One of the input signals is displayed during positive swings of the square wave, and the other is displayed during negative swings of the square wave. The chopping or switching signal is usually 50 kHz or less for electronic switches. The shortcoming of dual-trace display using an electronic switch is the inability to display any meaningful signal at a frequency above one-tenth the frequency of the chopping signal. For example, if the chopping frequency of an electronic switch is 50 kHz, any attempt to display a waveform whose frequency is above 5 kHz will result in a very distorted display. Other pitfalls of the electronic switch method are low bandpass, usually less than 300 kHz, and no calibration of input signals.

The only practical way to get an accurate, calibrated dual-trace display is to use dual vertical amplifiers to provide full dual-channel input. This method gives complete control over signal amplitudes, coupling, trace position, and display mode.

Plug-In Scopes

Most advanced laboratory scopes, as exemplified by the Tektronix 5103N in Fig. 8-9, are of the plug-in type. The plug-in type consists of a mainframe containing, at the least, the low-voltage and high-voltage power supplies, and the crt and its associated circuitry. With this type of mainframe, the entire vertical amplifier is a removable, modular unit. Similarly, the time base and the horizontal sweep amplifier comprise another removable, modular unit. These removable, modular units fit into compartments of the mainframe. The compartments are open in the front to receive the plug-in units, and have connectors at the back, which mate with connectors at the back of the plug-ins.

The Tektronix 561A is an example of the type of design in which the mainframe contains only the power supplies and crt circuitry. This scope is illustrated in Fig. 8-10. The holes in the front panel, through which the plug-in units are inserted into the compartments provided in the mainframe, are described by the broken lines on the front panel in the illustration. Notice that the hole on the left accepts the vertical module, and the hole on the right accepts the horizontal module.

Other types of mainframes may also contain the vertical deflection amplifier and the entire horizontal sweep system.

Tektronix scopes of the 550 series exemplify this type of design. In this type of scope, only a vertical preamplifier (preamp) is contained in a plug-in unit. A number of different plug-ins are available. The capabilities of the scope depend on the vertical preamp plug-in selected. If, for example, a dual-trace vertical preamp is selected, two signals may be displayed simultaneously. If, on the other hand, a four-trace

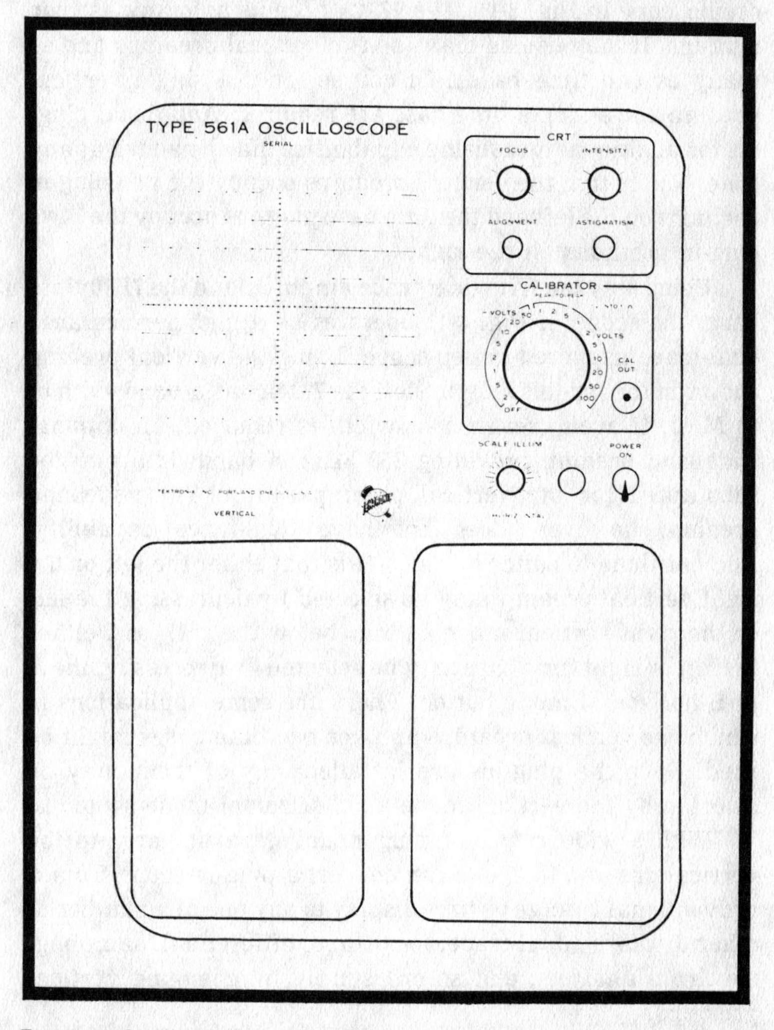

Fig. 8-10. An illustration of the Tektronix 561A scope showing the front panel openings of the vertical and horizontal plug-in compartments. (Courtesy Tektronix, Inc.)

vertical preamp is used, four signals may be displayed simultaneously.

Another type of mainframe contains the vertical amplifier, but only the amplifier part of the horizontal sweep circuit. In mainframes of this type, the vertical preamp is contained in one plug-in and the time base is contained in another plug-in. This is the design of the Tektronix 7704A oscilloscope in Fig. 8-11. The 7704A accepts as many as four plug-ins. It may use as many as two vertical preamps and as many as two time bases. Of course, only a single vertical preamp and a single time base are required. Additional plug-ins for additional measuring capabilities may be added at any time. Notice that the vertical preamps occupy the two plug-in positions on the left and the time base plug-ins occupy the two plug-in positions on the right.

Using only the 7A18 dual-trace amplifier and the 7B70 time base, the scope in Fig. 8-11 operates as a high-performance dual-trace triggered-sweep scope. Using the vertical preamp shown in the far-left plug-in slot, the 7704A has a bandwidth of 80 MHz. If even greater bandwidth is required, an optional wideband preamp providing 250 MHz of bandwidth may be substituted for the vertical preamp shown. The wideband preamp, however, does not have dual-trace capability. Another thing to notice in Fig. 8-11 is that either the left or the right vertical preamp may be selected by depressing the left or the right vertical mode button (below the crt), and either the left or right time base may be selected by depressing the A or B horizontal mode button. There are some applications in which two vertical preamps and / or two time bases might be used. Once the plug-ins are installed, any of them may be selected by the vertical mode and horizontal mode switches.

With a wide range of plug-in units, particularly in the vertical preamp line, one can convert a plug-in scope from a conventional voltage vs time display to any one of a number of other displays (dual-trace, four-trace, differential, sampling, spectrum analysis, and so on) simply by changing vertical preamps.

The chief advantage of plug-in scopes is their versatility. The advantages of plug-in scopes do not come cheap, however.

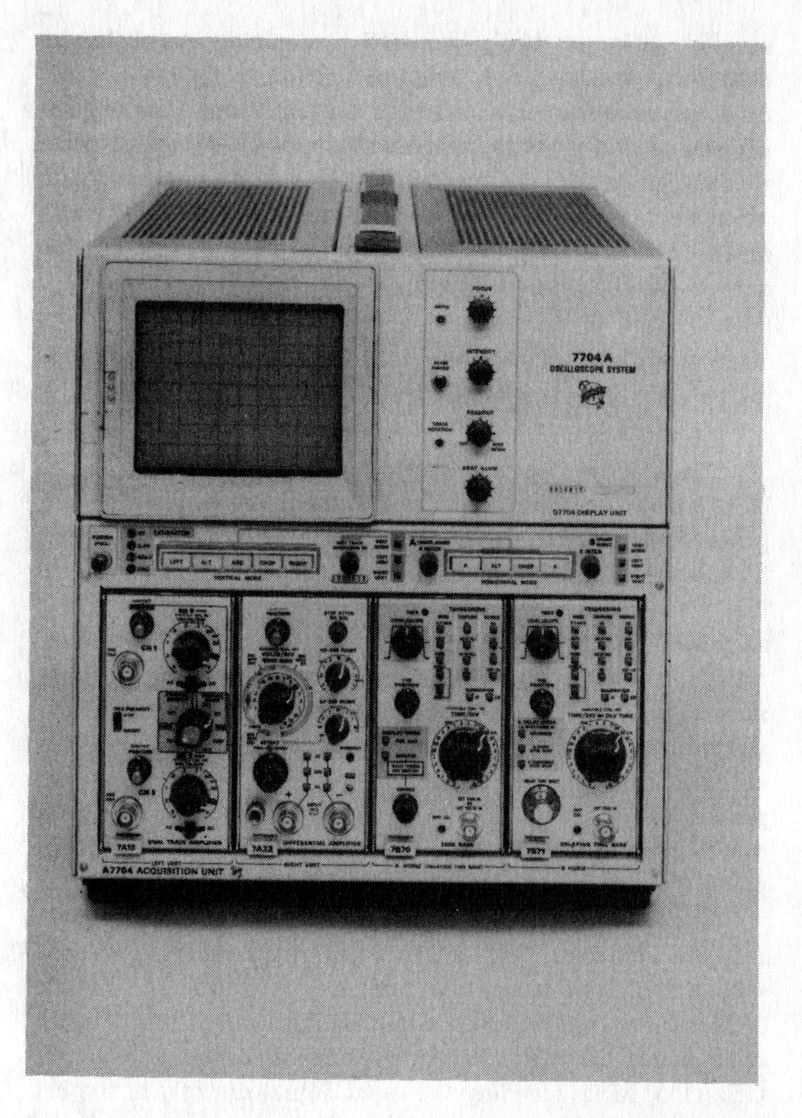

Fig. 8-11. The Tektronix 7704A accepts as many as two vertical preamps and two time-base plug-ins. The plug-ins used will depend on the application of the scope. (Courtesy Tektronix, Inc.)

The 7704A mainframe costs $2400, the dual-channel amplifier shown in Fig. 8-11 costs $535, and the 7B70 time base costs $625. Equal or better performance in any one measurement area can usually be obtained for a lower (but not greatly lower) cost in a non plug-in type. Of course when measurements in several different areas normally requiring separate instruments are required, nonplug-in units can have a great price advantage. Often, a single plug-in scope can replace several special-purpose instruments having a combined price much greater than the price of the plug-in instrument. Also, when the ultimate in scope performance is required, there is no alternative to the plug-in scope, since the top-of-the-line lab scopes use the plug-in design.

ULTRAWIDEBAND SCOPES

These scopes have a bandwidth capability of up to 500 MHz, and a rise time as low as 0.8 msec. The acquisition and processing of signals in the 500 MHz region require techniques very different from those used to handle signals in the 100 MHz region. In addition to special circuitry, ultrawideband scopes use a special deflection system in their crt. Following travelling-wave tube construction, these crt use an array of deflection plates so mounted and spaced that the electron beam, in passing between successive plates, receives an additional amount of deflection. This construction improves the effective bandwidth of the last driver stage in a manner similar to the way a distributed amplifier reduces parallel capacitance by using plate and grid delay lines.

The scope in Fig. 8-12 offers the ultimate in performance. It is the Tektronix 7904—a general-purpose laboratory scope with the highest bandwidth currently available. Using the 7A19 vertical amplifier plug-in, this instrument has a bandwidth of dc to 500 MHz. Since the bandwidth of the crt itself is 1 GHz (1000 MHz), feeding the input signal directly to the crt permits the display of signals having a frequency as great as 1 GHz. Using the 7A21N direct-access plug-in, a direct access connection to the crt is made, bypassing the vertical amplifier and disconnecting the second vertical channel. With the direct-access unit in use, the scope can display signals having a rise time as short as 0.35 nsec. Over 30 plug-ins are available for this scope, making virtually any measurement possible. We will discuss some of the available plug-ins in this chapter.

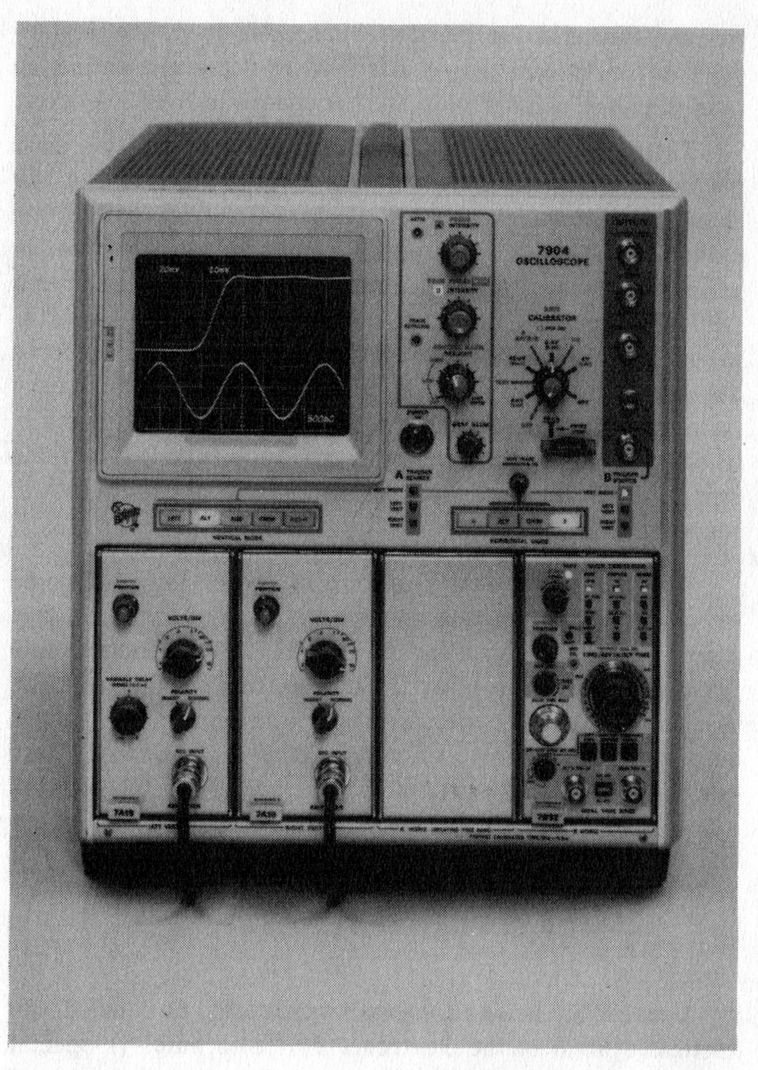

Fig. 8-12. The Tektronix 7904 scope has a bandwidth of dc to 500 MHz. This is the top-of-the-line model of the largest manufacturer of high-performance scopes. (Courtesy Tektronix, Inc.)

The major drawbacks of the helical-deflection crt used in this type of scope have been limited scan, low sensitivity and high cost. The problems of limited scan and low sensitivity have been overcome in the crt of the Tektronix 7904 by using a dome-shaped mesh electrode between the deflection-plate structure and the postaccelerator field. The mesh effectively screens the beam in the deflection area from the postaccelerator field, and shapes the field to achieve a deflection magnification of x2 in both the horizontal and vertical axes.

The optimum shape for the mesh to achieve good geometry was determined by using a computer to plot the shape of the fields produced by the mesh and the path of the cathode-ray beam through these fields. The equation producing the desired shape for the mesh was then fed into a numerical-control machine, which made the tool for producing the mesh. The elaborate manufacturing procedures and sophisticated design techniques required to achieve the bandwidth of the 7904 imply expensiveness. The 7904 mainframe costs $2900, and with the plug-ins shown in Fig. 8-12, the cost of the instrument rises to $5300. This is the ultimate in scopes.

The helical deflection system of the crt is constructed using a special method to achieve the 1 GHz bandwidth. It is also relatively expensive to produce. To intensify the brilliance of the crt display, an accelerating potential of 25 kV is used. This high accelerating potential reduces the vertical sensitivity of the crt somewhat. The dome-shaped mesh electrode between the deflection plate and the postaccelerator field, however, results in very good sensitivity.

Vertical

Because of the high frequencies that must be handled, the vertical system of the Tektronix 7904 also employs special design features. The interface between the mainframe and the plug-ins is especially designed to accommodate the ultrahigh frequencies. The characteristic impedance at the interface (mainframe to plug-in connection) is 50 ohms, an optimum value for transmitting UHF signals. The signal paths between circuit elements are all transmission lines, which are terminated in their characteristic impedance. This results in a very clean transient response, with normally less than 5 percent overshoot.

The delay line, required to permit viewing the leading edge of a triggering signal, also contributes to clean response. It is designed for maximum delay in minimum volume, and for short rise time. The line consists of two parallel solid conductors in a polyethylene dielectric with a foil wrap and a polyethylene protective jacket.

A number of different plug-in vertical preamps are available. The vertical preamp used determines the maximum bandwidth and some other characteristics of the scope. To realize the full bandwidth potential of the 7904, the 7A19 plug-in amplifier ($500) is required. Two such amplifiers are required for dual-trace displays. The 7A19 plug-in vertical preamp has an input impedance of 50 ohms. The input signal passes through a 50-ohm attenuator, providing deflection factors of 10 mV per div to 1V per div. Attenuator switching is accomplished **ahead** of the preamp except when the attenuator is in the 10 mV per div position. The inherent sensitivity of the vertical preamp is 20 mV per div. Since the 50-ohm line from the mainframe to the plug-in is terminated at both ends, switching out the termination at the source end increases the gain by a factor of two, giving the 10 mV per div sensitivity.

The first amplifier stage, illustrated in Fig. 8-13, is called an **ft** doubler. The schematic of this unique wideband amplifier is greatly simplified to provide insight into the

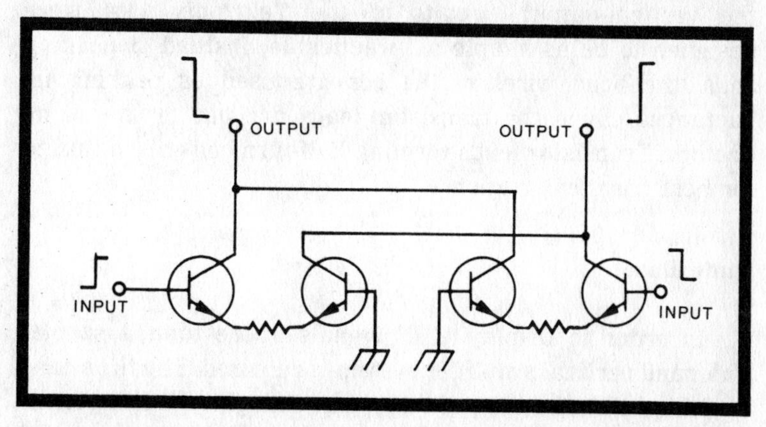

Fig. 8-13. A simplified schematic of the **ft** doubler circuit used in the vertical amplifier of the Tektronix 7904 scope. (Courtesy Tektronix, Inc.)

operation. The **ft** symbol represents the frequency at which the common-emitter current gain is unity. With ordinary cascading of stages, no more than unity current gain is possible. However, the **ft** amplifier achieves greater current gain by effectively placing the base-emitter inputs of four transistors in series. The push-pull arrangement permits the collectors of the transistors to e be, in effect, paralleled. At frequency **ft**, the current gain of the circuit is approximately 2, twice the current gain of an ordinary common-emitter circuit. Using several of these circuits in the vertical amplifier, an amplifier with appreciable **current**gain at **ft**can be built.

The vertical amplifier, in the mainframe, uses three **ft** doubler circuits with coupling between stages provided by 50-ohm transmission lines. Each **ft** stage is built on a separate integrated-circuit (IC) chip, using special high-frequency IC fabrication techniques. The resistors connecting the emitters are processed on the same strip with the transistors by depositing precise amounts of nickel and chromium on the IC substrate. Many other critical processes are involved in the construction of the high-frequency ICs.

The vertical-output amplifier is a hybrid IC with a substrate carrier on which are mounted five silicon chips. The five chips consists of an **ft** doubler, two chip capacitors, and two output transistors.

In order to minimize the attenuation of high frequencies, the vertical-output circuits of the Tektronix 7904 were designed to be as simple as practicable. Instead of peaking coils, the bond wires of the ICs are used as peaking inductances. Even the transistor leads are put to use as inductors. Transistor leads forming half-turn coils are adjusted for best transient response.

Time Base

In order to display UHF signals, more than a special wideband vertical amplifier system is required. The time base is also special. It must provide a fast enough sweep to permit viewing a reasonably small number of cycles of the vertical input signal. The 7B92 dual time-base plug-in used with the

7904 has a fast sweep, 0.5 nsec per cm, and has both alternate and chopped modes. The top sweep rate of 0.5 nsec per cm places stringent demands on the horizontal amplifier.

The horizontal deflection sensitivity of the crt is 7V per cm and the beam must be swept across the 10 cm width of the crt in 5 nsec when the horizontal sweep rate is at its fastest. That means the output amplifier must swing through 70V in just 5 nsec. Fig. 8-14 is a simplified schematic of the circuit used to provide the fast, large-signal amplification needed in the horizontal output amplifier.

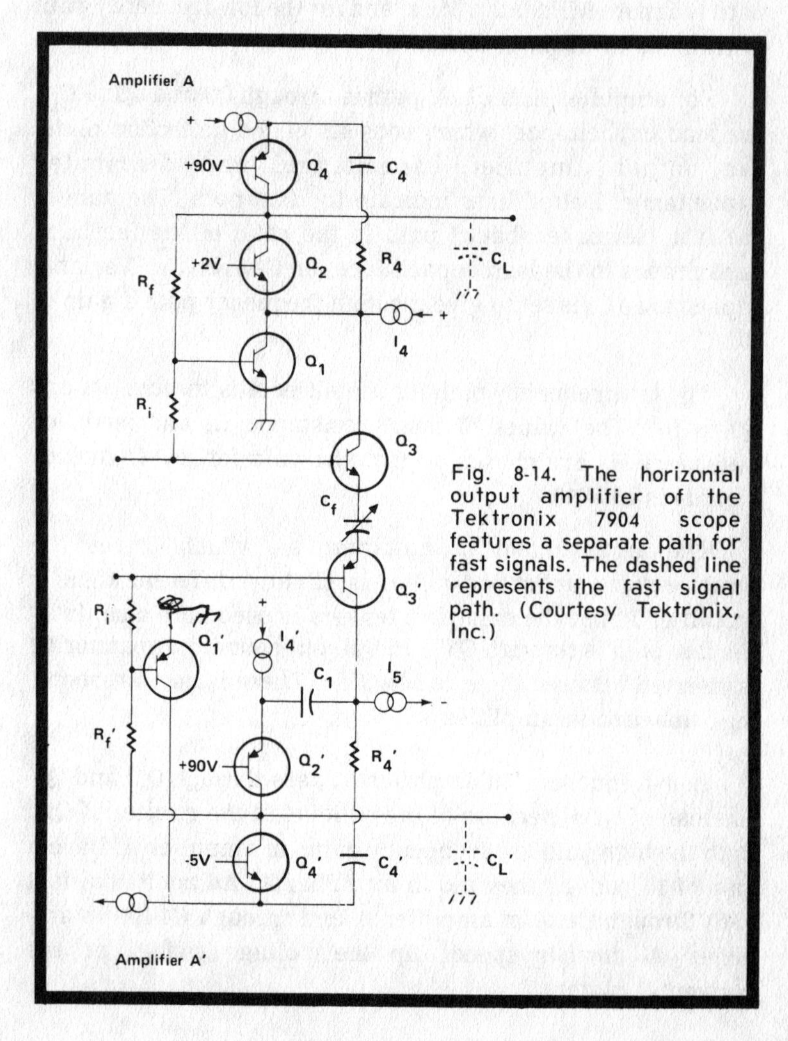

Fig. 8-14. The horizontal output amplifier of the Tektronix 7904 scope features a separate path for fast signals. The dashed line represents the fast signal path. (Courtesy Tektronix, Inc.)

The horizontal amplifier actually consists of two amplifiers, which are designated A and A' in the figure. Amplifier A drives the negative-going deflection plate, and is designed to have optimum performance in the negative direction of output. Amplifier A' drives the positive-going deflection plate and has good performance for positive-going output signals.

A unique feature of amplifiers A and A' is that each amplifier has two signal paths to the plate it drives—a high-frequency path using series feedback, and a low-frequency path using shunt feedback. Bandwidth for the high-frequency path is about 1 MHz to 200 MHz, and for the low-frequency path is from dc to about 30 MHz.

For amplifier A, the fast path is through Q3 and Q2 to C_L, the load capacitance, which consists of the deflection plate, the output amplifier capacitance, and distributed capacitance. Dotted lines indicate the fast path. The gain of the fast (series-feedback) path is the ratio of the feedback capacitance to the load capacitance, or C_f over C_L. Variable capacitance C_f is set to give the high-frequency path a gain of 10.

The low-frequency path for amplifier A is through Q1 and Q2 to C_L. The values of input resistance R_i and feedback resistance R_f are chosen so that the gain for low-frequency signals is also 10.

The arrangement of amplifier A', which drives the positive-going deflection plate, is slightly different than in amplifier A, but the dual-path feature is used here also. In A' the fast path is through Q3' and Q2', but coupling capacitor C1 is inserted between them to block dc. There is no corresponding capacitor in amplifier A.

Low-frequencies in amplifier A' pass through Q1' and Q4' (instead of Q2'), because of the dc level at the emitter of Q2'. Both the high- and low-frequency paths in amplifier A' have a gain of 10, just as they did in amplifier A. An additional fast path through C4R4 in amplifier A and through C4'R4' in amplifier A' further speeds up the voltage swings at the horizontal plates.

SWEEP MAGNIFICATION AND DELAY

Sometimes we want to display parts of waveforms that occur long after a suitable triggering signal is available. Such signals can be displayed if a long enough sweep is available. The sweep must be as long as the time between the triggering signal and the signal to be observed plus the time period of the signal to be observed. For example, if the time between the trigger signal and the waveform portion to be displayed is 50 usec and the time period of the waveform portion to be displayed is 10 usec, the sweep length must be greater than 60 usec, as illustrated in Fig. 8-15. Notice that the portion of the wave we are interested in occupies one-sixth of the screen width. If the portion we were interested in were narrower still, it would occupy an even smaller part of the screen. If the duration of the desired waveform is too short compared to the duration of a full sweep, an accurate examination may not be possible. In this case, we need in some way to magnify, or expand, the display for the time during which the event we wish to see occurs. There are two ways of doing this.

The inexpensive and simple way of doing this is to increase the gain of the horizontal amplifier, allowing one end of both ends of the display to go off-screen. Then, use the horizontal positioning control until the desired portion of the

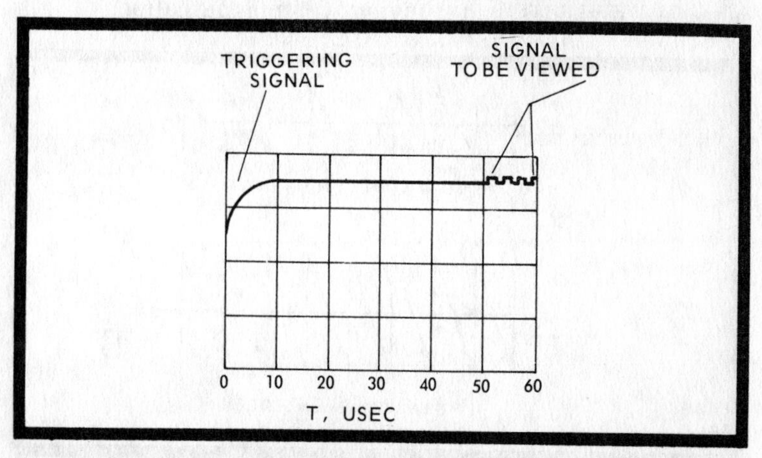

Fig. 8-15. When there is a relatively long interval between the triggering signal and the signal to be viewed, the signal to be viewed will occupy but a small portion of the screen.

display is on-screen, if it happens to be off-screen. This method delays the **presentation** of the sweep portion we see. The sweep still begins with the triggering signal; turning up the horizontal gain doesn't change that—but the electron beam comes on-screen later. The voltage on the positive horizontal deflection plate rises for some time before the beam comes on-screen. Thus, the time at which the beam sweeps across the screen is delayed.

As Fig. 8-16 shows, turning up the horizontal gain increases the amplitude of the sawtooth sweep voltage, but not the period of the sawtooth. Since the voltage change on the horizontal plates during a sweep is greater and the time during which the change occurs is unchanged, the voltage must rise at a faster rate than before the horizontal gain was turned up. This means that the electron beam will sweep across the screen faster, producing display magnification.

Some scopes have a switch for turning up the horizontal gain a definite, fixed amount to produce x5 or x10 magnification. When the control is activated, the sweep rate is raised to 5 or 10 times the setting of the horizontal sweep switch, or time-per-division switch. As shown in Fig. 8-17, the center division of the unmagnified display is the portion visible in the magnified display. The center portion of the waveform is thus displayed the way it would look if the screen were 100 divisions wide, giving x10 magnification.

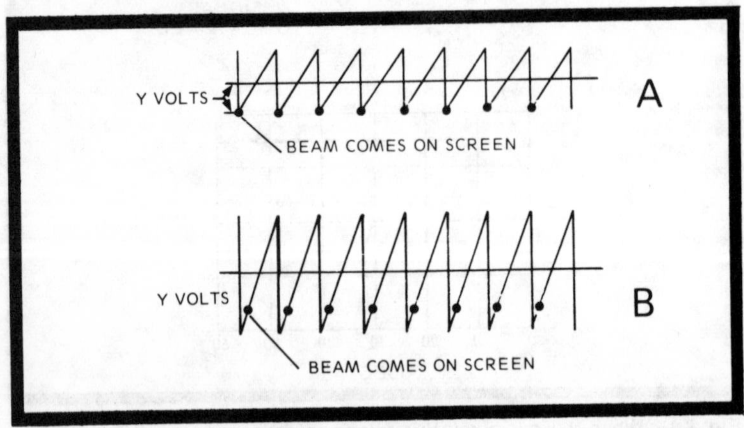

Fig. 8-16. Turning up the horizontal gain increases the amplitude of the sawtooth, but leaves its period unchanged.

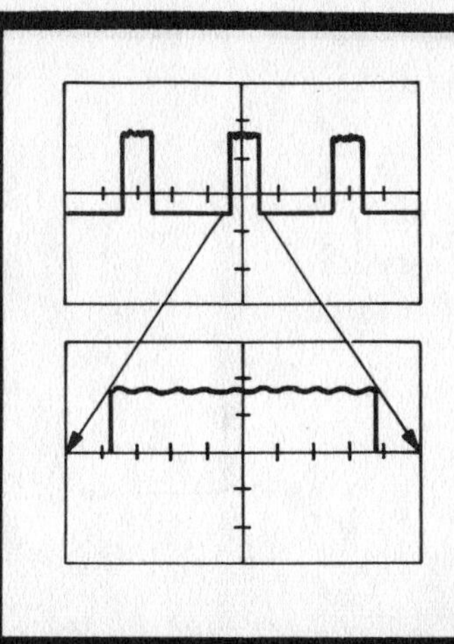

Fig. 8-17. With x10 magnification, a 1-division unmagnified portion of a display can be expanded to cover ten divisions.

Another way of producing a magnified display is through the use of **delaying sweep**. This method, which we will explain, has several advantages:

1. Greater effective magnification.
2. Elimination of jitter and drift in the displayed waveforms.
3. Greater accuracy of time measurements between events in the waveforms.
4. Better long-term accuracy of the displayed time base.

Delaying sweep measurements requires the use of two linear time bases. The two time bases required to make delaying sweep measurements with the Tektronix 7704A scope are shown in Fig. 8-11 inserted into the horizontal compartments of the mainframe. Notice that a special delaying time base is used for this measurement and that it looks like the regular time base except for a delay-time multiplier, or DTM, control. This time base may be considered the main time base, and we will refer to it as time base A. The delaying time base allows the scope operator to select a specific delay time. When this time is reached, the sweep of the other time base, the delayed-sweep time base, begins. We shall call this other time base, time base B. Time base B is typically 10 or 100 times faster than time base A. The combination of sweeps A and B provides for display magnification and accurate time-interval measurement. Let's see how.

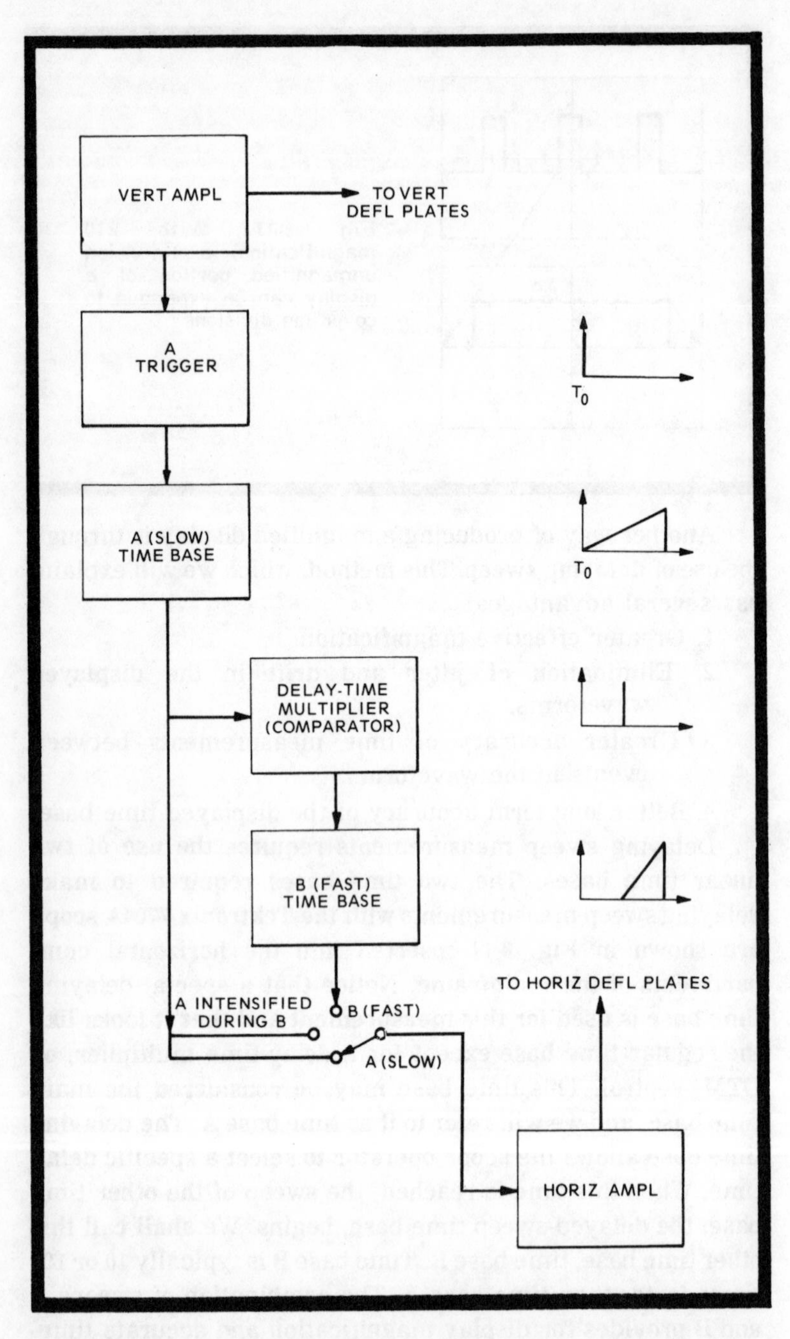

Fig. 8-18. A block diagram illustrating the principle of delayed sweep.

In Fig. 8-18 you see a block diagram illustrating how the delayed sweep is produced. The waveform diagrams in the figure represent the output of the blocks immediately to the left of them. At t = 0 on the waveform diagrams, the initial trigger, that is, the A trigger, occurs. This is obtained from the signal applied to the vertical amplifier. With the horizontal display switch in the A sweep position, single time-base operation is obtained as described at the beginning of Chapter 4. However, the A-sweep voltage ramp is applied to a voltage comparator, which produces a trigger pulse at a later time, t', at which time the A-sweep voltage has risen to some predetermined level. The trigger level may be set to occur anywhere along the A-sweep ramp. The comparator is set to deliver a trigger signal midway during the A sweep, to start the B sweep. The time between t_0 and t', that is, the time between the start of the A sweep and the start of the B sweep, is the time delay.

The time delay is adjusted by means of the delay-time multiplier control of the comparator. This DTM control has a 10-turn dial that reads between 0 and 1.000. If the DTM were set at 1.000, the B sweep would begin at the end of the A sweep. If the DTM were set at 0, the B sweep would begin with the A sweep. In the illustration, it must be set at 0.500, since the B sweep begins halfway through the A sweep. If the A sweep rate is 50 usec per cm and the screen width is 10 cm, the A-sweep time is 500 usec. If the DTM is set at 0.500, the B sweep begins at the halfway point of the A sweep, that is, at 250 usec after sweep A starts. Thus, we see that the delay is found by multiplying the A-sweep time by the DTM setting. An easier way is to multiply the A sweep-rate setting by 10 cm (the width of the horizontal sweep), and then to multiply the product of this multiplication by the DTM setting. In the case just given, the delay would be calculated thus:

$$\text{Delay} \doteq 10\ \text{cm} \times \frac{50\ \text{usec}}{\text{cm}} \times 0.500 = 250\ \text{usec}$$

As another example, if the A sweep is set at 100 usec per cm and the delay-time multiplier dial reads 0.350, the delay is 350 usec.

To summarize, the A trigger starts the A sweep. When the A sweep voltage rises to some definite value, as determined by the DTM control of the comparator, the comparator delivers a trigger pulse to the B time base, starting it up. The time between the start of the A sweep and the start of the B sweep is the delay time. The part of the A sweep accomplished before the B sweep starts is determined by the DTM setting. The length of time represented by this portion in turn depends on the sweep rate of the A time base. Thus, the delay time depends on the DTM and A-sweep settings, and can be found by multiplying them together and then multiplying by the screen width. The B, or delayed, sweep is typically 10 or 100 times faster than the A sweep.

If the horizontal display switch is set at the "A intensified" position, the horizontal deflection is still controlled by the A time base. With this setting, additional circuitry not shown in Fig. 8-18 is brought into play to intensify the trace during the time of the B sweep. Inasmuch as the intensified portion occurs only during the B sweep, its starting point is set by the DTM dial and its length by the B sweep time.

Finally, magnification occurs when the horizontal display switch is set at the B sweep position. In this position, the B sweep is the one applied to the crt via the horizontal amplifier, and since it is a faster sweep than the A sweep, magnification results.

Displays obtained with three different settings of the horizontal display switch are shown in Fig. 8-19. The display at the top was obtained with an integrated square wave applied to the vertical input, and with the A time base providing the horizontal sweep at a rate of 50 usec per div. This is the same display as would be obtained with the delayed-sweep feature inoperative.

The center display was obtained with the horizontal display switch set at the "A intensified during B" position. Again, the A time base is providing the horizontal sweep at a rate of 50 usec per div. The B sweep begins after the A sweep is about 55 percent completed, intensifying the trace as long as the B sweep lasts. The DTM setting is 0.550, thus, the delay must be 10 x 0.550 x 50 usec, or 275 usec. Since the A sweep rate is 50 usec per div and the B sweep occupies about one division of the A sweep, the B sweep time must be 50 usec. As shown by the caption in the figure, the B sweep rate (bottom display) is 5 usec per div.

When the horizontal display switch is set in the "B sweep" position, the B time base instead of the A time base provides the horizontal sweep. The intensified portion of the center display is displayed full-screen in the bottom display. Since the B sweep has taken over, the B sweep rate is used for making time measurements. Every division of the time axis now represents 5 usec. The time required to sweep across all 10 divisions is 10 times this, or 50 usec, as stated earlier.

The delayed sweep may be used to make measurements with errors as low as 1 percent between two points on the delayed sweep waveform. To do this, the first point is set at

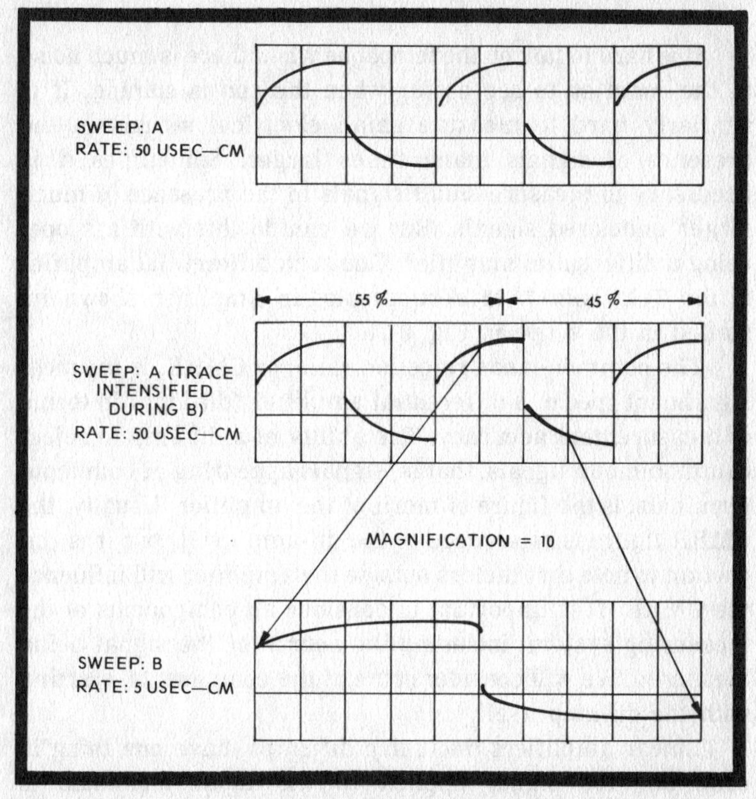

Fig. 8-19. Displays obtainable with a delayed-sweep scope.

one of the vertical lines on the graticule and the DTM reading noted. Then, the second point on the waveform is set at the same vertical line on the graticule by adjusting the DTM. The time between the two points on the waveform is the difference between the DTM settings, times the A-sweep setting, times the length of the time axis.

t (time interval) = (DTM / DTM') x time / div of A x div.

The last factor in the above formula represents the number of divisions (cm or inches) in the horizontal axis, 10 div for the scopes in Figs. 8-9 through 8-12.

DIFFERENTIAL AMPLIFIERS

It is hard to talk on the telephone when there is much noise in the room or to see a star when the sun is shining. It is similarly hard to measure small electrical signals in the presence of signals many times larger. Sometimes it is necessary to measure small signals in the presence of much larger undesired signals. But we can do this with a scope, using a differential amplifier. One such differential amplifier is the Tektronix 7A22 vertical plug-in amplifier shown installed in the scope in Fig. 8-11.

The common-mode rejection ratio, or CMRR, is the most significant spec of a differential amplifier (dif-amp) in terms of measurement accuracy. The ability of a dif-amp to reject common-mode signals, that is, signals appearing at both input terminals, is the figure of merit of the amplifier. Usually, the CMRR figure is referenced to the dif-amp itself, but it is important to note that factors outside the amplifier will influence the CMRR. It is important to consider all components of the measuring system, including the source of the signal being measured. We will consider some of the components, starting with the dif-amp itself.

Linear amplifiers, including dif-amps, have one thing in common: the signal applied to the input terminals is multiplied "gain times." That is, an amplifier will amplify "gain times the difference between the input terminals." This leads to two important facts:

1. The amplifier output voltage E_o can be expressed as

$$E_o = A_v (E_i1 / E_i2)$$

where E_i1 and E_i2 are the input terminal voltages.

2. The gain A_v of an amplifier is fixed, and does not depend on external factors.

A dif-amp is a double-ended amplifier of the push-pull or paraphase type, which amplifies the **difference** between two signals applied at its input terminals. If the two signals are identical, there is no difference to be amplified, and hence no output. This is illustrated in Fig. 8-20. Assume that amplifiers A1 and A2 each have a gain of 10. As we said, this gain is fixed. In the amplifier system shown, the gain is measured from side to side, i.e., from the output of A1 to the output of A2. The gain to be measured between these points should be "gain times the difference between the input terminals." The difference between the input terminals is a difference in amplitude or phase, or both. It may be described as the graphical addition of the two signals appearing at the input terminals of both amplifiers. Referring to signal (a) in the figure, we can see two equal input voltages of opposite polarity, which result in an output voltage that is 10 times their difference. This is calculated to be 20V, and is represented graphically at the right side of the figure.

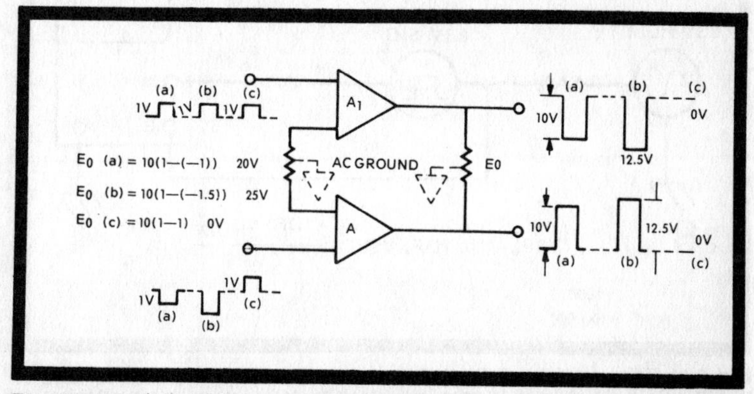

Fig. 8-20. A differential amplifier amplifies the difference between two signals.

Looking at signal (b), we see two input signals of opposite polarity but unequal amplitude. The output resulting from these voltages is calculated to be 25V. Notice that there is a minus sign attached to the 1.5V signal, since its polarity is opposite that of the 1V signal. Subtracting minus 1.5 from 1 gives 2.5, and gain (10) times this gives 25V, the output voltage. Note that while the input signal was unbalanced, the output signal is balanced—12.5V between output A1 and ac ground, and 12.5V between output A2 and ac ground. This shows how push-pull amplifiers tend to correct unbalanced signal drives.

Finally, looking at signal (c), we have two input voltages that are equal in both amplitude and in phase. Since there is no

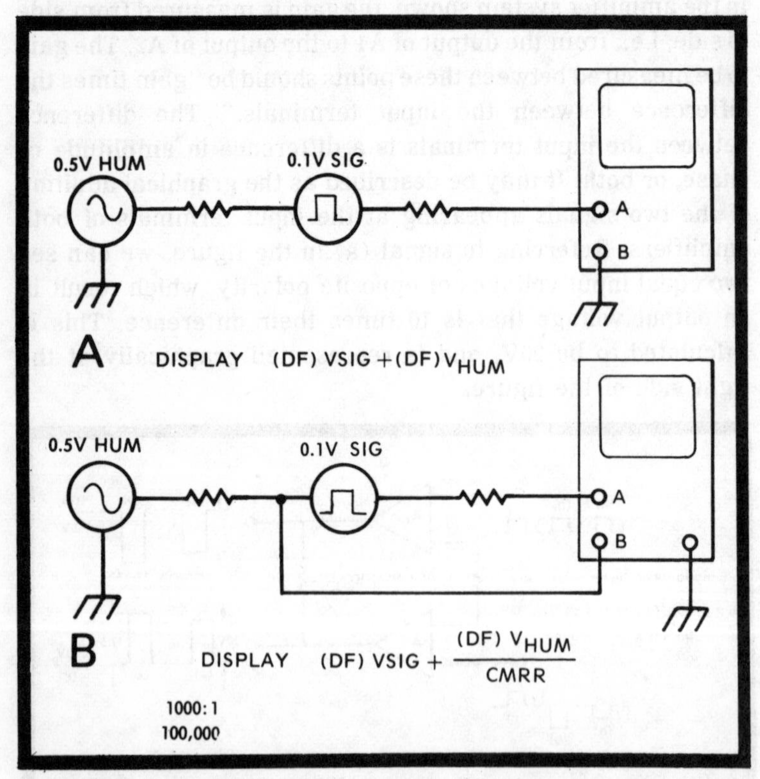

Fig. 8-21. The use of the differential mode in B reduces the hum voltage in the display by a factor of 1000, the CMRR. (In both A and B, CMRR is 1000:1 and V per div is 0.05.)

difference signal developed across the input, there is no signal developed at the output. Input signals such as signal (c), which are the same at each input terminal, are termed common-mode signals. These are the signals rejected by dif-amps.

Fig. 8-21 shows a measurement problem that can be solved with a dif-amp. The operator wants to observe the output of a signal source and connects his scope in the conventional manner illustrated in Fig. 8-21. Now the desired signal is a square wave of 0.1V, and the interfering signal is a hum voltage picked up from the power line and having a value of 0.5V. With the connection shown, these signals **add**. The desired signal represents 2 divisions of vertical deflection with the deflection factor set at 0.05V per div; the 5-times-larger hum voltage represents 10 divisions. Thus, the hum voltage effectively masks the desired signal, making measurement of the desired source difficult or impossible.

Using the differential scope mode in Fig. 8-21B, the problem can be licked. The signal voltage develops across inputs A and B as a 2 div input, as before. Hum develops as 10 div of **common-mode** signal. It is a common-mode signal because it is in phase at inputs A and B. If the dif-amp were perfect, theory says that the hum signal would be rejected completely. However, practical dif-amps are imperfect, so cannot completely reject a common-mode signal. The extent to which they do reject a common-mode signal is the CMRR previously mentioned. In the scope of Fig. 8-21, the CMRR is 1000: 1. Thus, the common-mode signal, the hum signal, is attenuated by a factor of 1000. That is, it will have an amplitude of only 0.01 div in Fig. 8-21B.

Whether or not a dif-amp will reject a common-mode signal depends on two things—the electrical parameters of the active devices in the amplifier, and the impedances over which the desired signal and the common-mode signal develop with respect to ac ground points.

The desired and common-mode signals see two different grounds in a dif-amp such as the one in Fig. 8-22A. The desired signal sees an ac ground between the two emitters, since the input signals to a push-pull amplifier cancel at this point, producing a zero reference point. The common-mode signal, on the other hand, recognizes only the actual chassis ground, the $-V_{CC}$ point.

R3 is the equivalent of two 20K resistors in parallel, as shown in the equivalent circuit, Fig. 8-22B. The common-mode signal will be divided between these two resistors, while the desired (push-pull) signal will bypass the resistors. Thus, the gain in the common-mode signal will be reduced by losses across the 20K resistors, while the gain in the desired signal will be undiminished. The gain in the signal might be calculated as follows, using typical values:

$$A_V = \frac{R1 + R2}{R_t1 + R_{t2}} = \frac{1000 + 1000}{10.2 + 10.2} = 98$$

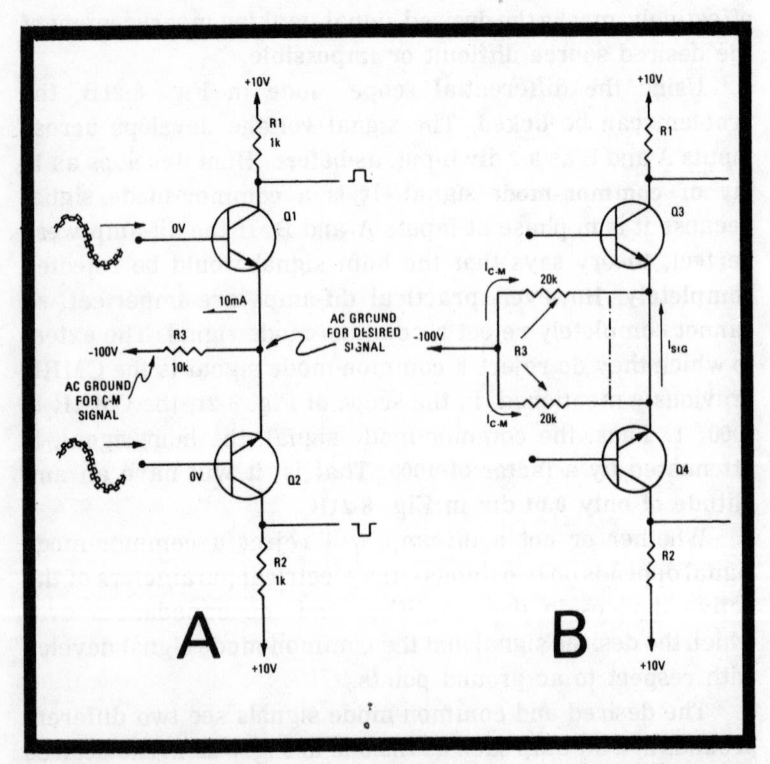

Fig. 8-22. In a dif-amp, the desired and the common-mode signals see two different ac grounds (sketch A). Sketch B shows the equivalent circuit and signal paths.

where R_t is the dynamic emitter resistance, plus the quotient of base spreading resistance and beta.

The common-mode gain in the hum would then be

$$Av = \frac{R1 + R2}{R_{t1} + R_{t2} + 2R3} = \frac{1000 + 1000}{10.2 + 10.2 + 2(10,000)} = 0.01-$$

Notice what the factor R3 does to common-mode gain. The inclusion of this factor in common-mode gain and the deletion of it from signal gain is explained from the different signal paths described in Fig. 8-22 and resulting from separate common-mode and signal grounds.

CMRR for **scopes** is the ratio of the deflection factor for a common-mode signal to the deflection factor for a differential signal. Thus, if we measure 10V per div across the input of a dif-amp and 1 mV per div across its output,

$$CMRR = \frac{10V}{10^{-3}} = 10,000:1$$

The dif-amp in Fig. 8-22A is current-driven by a technique known as "longtailing." The longtail in that circuit is the 10K resistor, R3, which supplies the current for both transistors. Generally, an amplifier current source approaching the ideal current generator allows high CMRR—for example, an infinitely high-impedance longtail returned to an infinitely high supply voltage. The constancy of the current source is one of the most important considerations in dif-amp design. Another important consideration is balancing of the two stages of the push-pull amplifier. Both stages should ideally have the same phase shift and amplification.

Component tolerances, especially in input attenuators, are the main cause of common-mode rejection (CMR) degradation. (CMRR is **not** CMR.) In some dif-amps (as the Tektronix 7A22 in Fig. 8-11), the input attenuators are not used in the most sensitive vertical ranges. In such cases, attenuation is accomplished by switching the gain in a feedback amplifier. In some other amplifiers, conventional attenuator methods are used, with attenuator components carefully selected and matched.

The CMR specs of a dif-amp are fixed, and can be assumed to be within the limits prescribed by the manufacturer. Most problems involved in differential measurements start with the methods used to connect the dif-amp to the measurement source.

Probes and interconnecting cables always introduce some degradation of CMR. The reduced CMRR due to the effect of probes and cables is termed **apparent CMRR**. Identical probes must be used for differential measurements, and special probes that can be optimized for CMR should be used. The effect of a difference between probes is illustrated in Fig. 8-23. If 0.5 percent resistors are used, the difference between the two resistors can be as great as 1 percent, and the CMRR in this case will be limited to 1111:1. Notice that if 5 percent resistors are used, the CMRR is limited to 111:1.

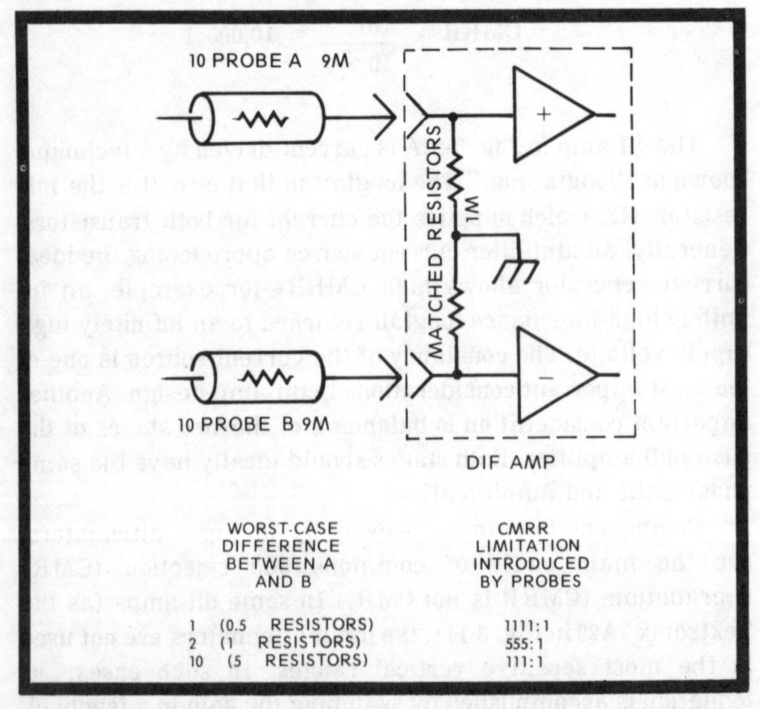

Fig. 8-23. The effect of the difference between probes on the CMR of a differential measuring system.

Impedance imbalance in the measuring equipment is not the only cause of CMR degradation. If the **source impedances** are different, as they usually are, the apparent CMR will be reduced even though the voltages from both sources are the same. This is illustrated in Fig. 8-24. In Fig. 8-24A, two 10.0000V sources designated A and B feed a dif-amp. Source A has an impedance of 100 ohms, source B, 50 ohms. Because the source impedances are different, the voltages at terminals A

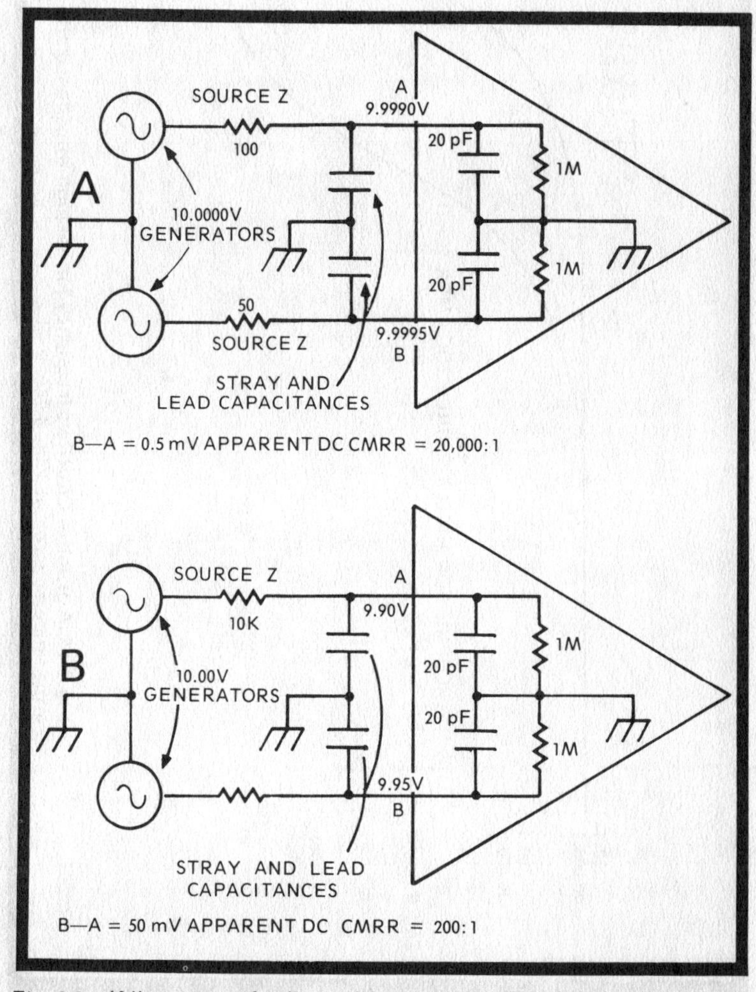

Fig. 8-24. If the sources feeding a dif-amp have different impedances, the CMRR will be degraded, especially if the sources are high-impedance sources. See Fig. 8-25 for effect on ac CMRR.

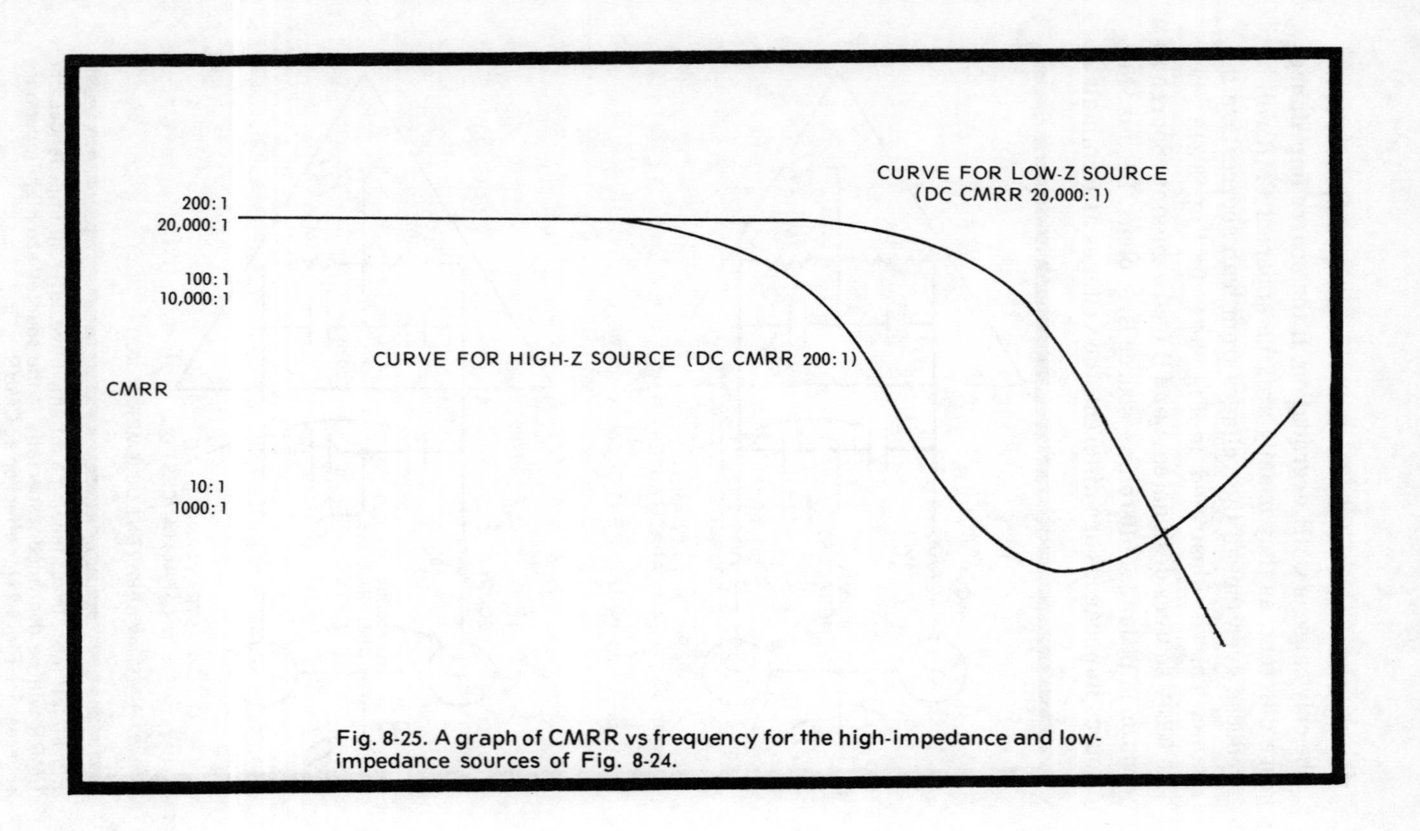

Fig. 8-25. A graph of CMRR vs frequency for the high-impedance and low-impedance sources of Fig. 8-24.

and B of the differential amplifier are different. One is 9.9990V and the other is 9.9995V, for a difference of 0.0005V, or 0.5 mV. The apparent CMRR resulting from this is graphed in Fig. 8-25. Notice that the dc CMRR for low frequencies is 20,000:1.

In the case just described, the degradation of CMR is not too bad. A different situation obtains in the case where the source impedances are relatively high, as in Fig. 8-24B. There the A and B source impedances are 10K and 5K, respectively. The impedance difference results in a differential input difference of 9.95V minus 9.90V, or 0.05V (50 mV). Fig. 8-25 shows the effect on CMRR. Note that at low frequencies the CMRR is only 200:1 for the high-impedance sources. **Also note the way CMRR drops off with increasing frequency.**

A dif-amp, because it **compares** two signals and amplifies the difference, is sometimes called a **comparator**. This can be misleading as the term **comparator** is more properly used to refer to differential units used for making voltage measurements with a scope. A differential comparator consists of a dif-amp **and** an accurately calibrated adjustable voltage source. When making voltage measurements with this apparatus, the calibrated dc comparison voltage is internally applied to offset any unwanted portion of the applied signal. This permits the measurement of small signals riding on large signals.

Signal voltages to be monitored frequently vary about a dc level. The dc will, of course, be blocked by an input capacitor if ac coupling is selected; but low-frequency components distort the least when dc coupling is used. If dc coupling is selected and the ac variations riding on a dc level are small compared to the dc level, the variations will appear on the crt as small deflections. Using a comparison voltage, with differential amplification, you're able to select a more convenient deflection factor.

Suppose, as an example, you wished to observe 20 mV ripple on a 100V power supply output and that you needed to direct couple to observe all ripple frequency components. Suppose further that your scope has a 1 mV per div deflection factor and an 8 div vertical scale. In order to keep the dc level on-screen, an attenuation factor of 10,000 would be required.

Using this attenuation factor, you might calibrate the scope for, say, 20V per div. However, the 20 mV ripple would represent only one-thousandth of a division—that is, 0.001 cm of vertical deflection!

Obviously, using conventional methods, the ripple would be too small to observe. Using the dif-amp, however, the ripple could be measured. You could monitor the power supply on either channel A or channel B, of the dif-amp, using x10 attenuation. This would give a calibration of 10 mV per div. Next, you could inject 100 **comparison** volts into the B channel input amplifier (or A channel input amplifier if the power supply is connected to B). The dif-amp would amplify only the signal differences. The 100V applied to the B input would cancel the 100V power applied at input A, **and only the 20 mV ripple would be amplified.** Since the scope was calibrated for 10 mV per div, the 20 mV ripple signal would deflect two divisions, enough to observe it and measure it.

Using the differential comparator, the scope becomes a precision null-seeking voltmeter. It can measure differential signals of only a few millivolts with an error of 0.25 percent or better. This accuracy closely approaches that obtainable with digital techniques!

STORAGE SCOPES

A storage scope is one that stores a waveform on its tube face after the waveform ceases to exist. The storage time may be for a few seconds or for hours. The stored display may be erased at any time desired to make way for a new display. Also, for measurements for which the storage mode of operation is not desired, the storage scope may be operated as a conventional, nonstorage instrument.

There a number of applications in which the storage facility is very useful. For viewing slowly changing phenomena that would appear only as a slowly moving dot on the crt screen, the storage facility is a must. Storage is also useful for viewing nonrepeating phenomena. If a phenomenon occurs only once, it can be observed only during the single occurrence with a conventional scope, and prolonged detailed examination of the signal is impossible. With the storage feature at one's disposal, it is possible to store the waveform of the nonrepeating phenomenon on the screen and view it at leisure. The storage scope is also useful for viewing recurring signals having a low repetition rate. If the repetition rate of a waveform on a conventional scope screen is less than about 30 times per second, the display flickers. Since a storage scope can store waveforms for hours, there is no noticeable fading of the display in the intervals between the occurrences of such waveforms. Another advantage of the storage feature is that is facilitates taking waveform pictures by permitting you to compose the picture. Unsatisfactory displays can be erased as many times as necessary before the photograph is actually taken.

In terms of design, the main difference between a storage scope and a conventional one is the special storage crt. It is the special crt rather than circuitry that gives the storage scope its ability to "freeze" waveforms. The vertical and horizontal amplifiers and time bases in storage scopes are no different than those in other scopes. About the only circuit differences are those to be found in the circuitry directly associated with the crt. There are two storage crt types: the **bistable** and the **half-tone** (variable-persistence).

Bistable Storage CRT

Fig. 8-26 depicts a bistable storage crt. The basic structure is the same as that of a conventional crt. However, there are several additional elements in a storage crt to make storage possible. The most important of these are the **flood guns**, which spray the entire screen with low-velocity electrons. The target of the guns in a storage crt consists of a phosphor screen with a thin, transparent, conductive coating in front of it.

Bistable storage is based on the principle of secondary emission. When a stream of electrons strikes a phosphor, secondary electrons are knocked off the phosphor surface. As the accelerating potential of the bombarding primary electrons is increased, more secondary electrons are knocked loose for each primary electron that strikes the screen. The ratio between the primary (bombarding) and secondary (displaced) electrons is called the secondary-emission ratio. When this ratio is less than one, the target area gains electrons and a negative charge. When the ratio is greater than one (more electrons leave the screen than arrive), the screen acquires a positive charge. The storage target has two stable potentials, two potentials it can be placed at to give a secondary-emission ratio of one. One is the voltage near the flood-gun cathode potential, and the other is near the collector potential.

When the storage scope is placed in the storage mode, the flood guns flood the screen with low-velocity electrons the way a floodlight illuminates a yard. This causes a slight background glow over the entire phosphor area. The flood electrons strike the phosphor, but do not have enough velocity to dislodge many secondary electrons. Consequently, the target area charges negative to the first stable state, that is, to a value near the flood-gun potential.

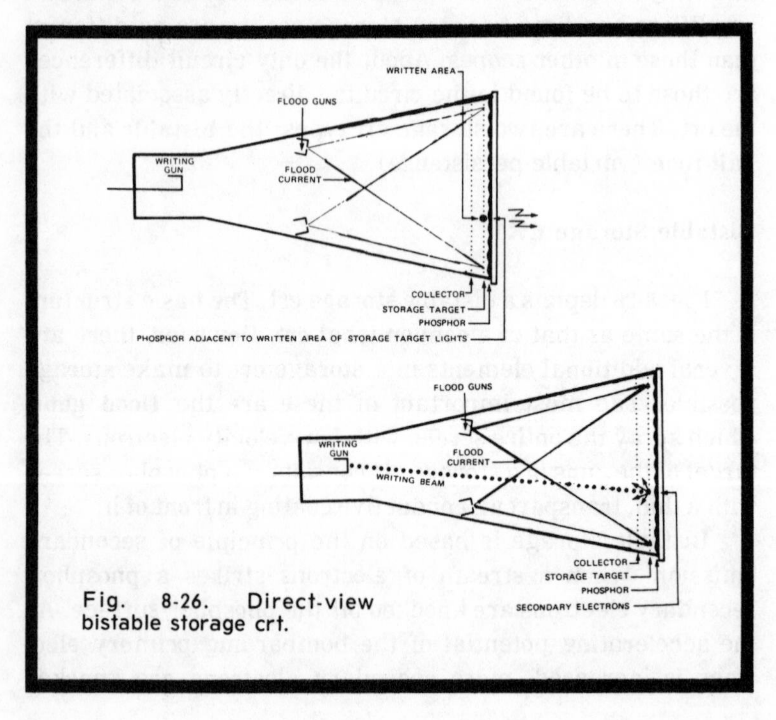

Fig. 8-26. Direct-view bistable storage crt.

As in any crt, the waveform to be displayed is traced out on the screen by a beam of high-velocity electrons. As this writing beam scans the screen in a storage crt, it dislodges many secondary electrons, because it consists of high-velocity electrons. The area traced over by the writing beam, the written area, flows and charges positive, while the unwritten area of the target shifts to the second stable state. Thus, we have on the screen in Fig. 8-26A a pattern of positive charges corresponding to the wave shape, against the negative background of the screen as a whole, which is at the collector potential. The situation is analogous to the paper of this page being negative and the print being positive.

The low-velocity electrons of the flood guns are now accelerated to the positively charged, written areas with enough velocity to dislodge more electrons than they add, and enough velocity to keep lit the phosphor areas just lit by the writing beam (Fig. 8-26B). As the flood-guns continue to bombard the written areas with accelerated flood-gun electrons, the written areas will remain positive and lit. Notice that storage in the storage crt results not from extraordinary phosphor persistence, but from a continuing flood-electron bombardment of the phosphor along the pattern that was traced out previously by the writing beam.

As the sweep rate of the writing beam increases, a maximum limit is reached where it is difficult to store a display. This limit, a spec for a storage scope, is known as **stored writing speed**. There are two techniques for extending the stored writing speed limit. For repetitive sweeps, the technique known as **integration** can be used. In this technique, the flood guns are momentarily turned off and repetitive sweeps are allowed to build up the charge on the target. This is like tracing over and over along the same path on a piece of paper—the pattern traced out stands out more each time it is traced out. In the storage crt, the charge built up over the writing beam path is allowed to build up until the written area of the target switches to the stored state. Then the flood guns are turned on to view the display. The electrons from the flood guns will continue to dislodge enough secondary electrons to keep the written area positively charged and will strike the phosphor with sufficient velocity to light it.

Another technique of increasing writing speed is called **enhancement**. This technique is used with fast, single-sweep displays only. When fast signals are being viewed, the writing beam may not shift the target area positive enough to store the trace. To aid in storing the partially written trace, an **enhance pulse** is applied to the target area during the sweep time. This pulse raises the target positive enough to shift the partially written trace into the stored state.

A stored display can be erased from a storage crt by first raising the collector to a positive voltage so that the entire screen is fully written. Then the collector is dropped to a negative potential about the same as the potential of the flood-gun cathodes, and finally it is slowly returned to the positive ready-to-write level.

A good example of the bistable storage scope is the Tektronix 7313 in Fig. 8-27. This scope features split-screen bistable operation as well as conventional operation. The split-screen storage crt of the 7313 provides the convenience of storage and conventional displays on the same crt **at the same time**. This capability is useful in many applications. For example, the operator may wish to store a reference trace and then view the change in waveform characteristics as he varies the values of circuit components. He can do this by operating half of the display in a stored mode and the other half in a conventional mode. Thus, amplitude, duration, and other characteristics of waveforms displayed in a conventional mode can be adjusted precisely to the stored reference trace. If you look closely at Fig. 8-27 you can see the separate controls for the upper and lower halves of the split screen.

The Tektronix 7313 has a stored writing speed of 5 div per usec. Stored traces can be viewed for up to 4 hours. Some storage crt require precautions to prevent burning the screen. The 7313 crt is extremely burn-resistant, and requires no special operating precautions.

The use of most of the special controls in the upper right part of the front panel of the Tektronix 7313 should be fairly apparent from previous discussion. There is another control that needs to be explained—the **auto erase** control. When the auto erase mode is selected by pressing the auto erase button,

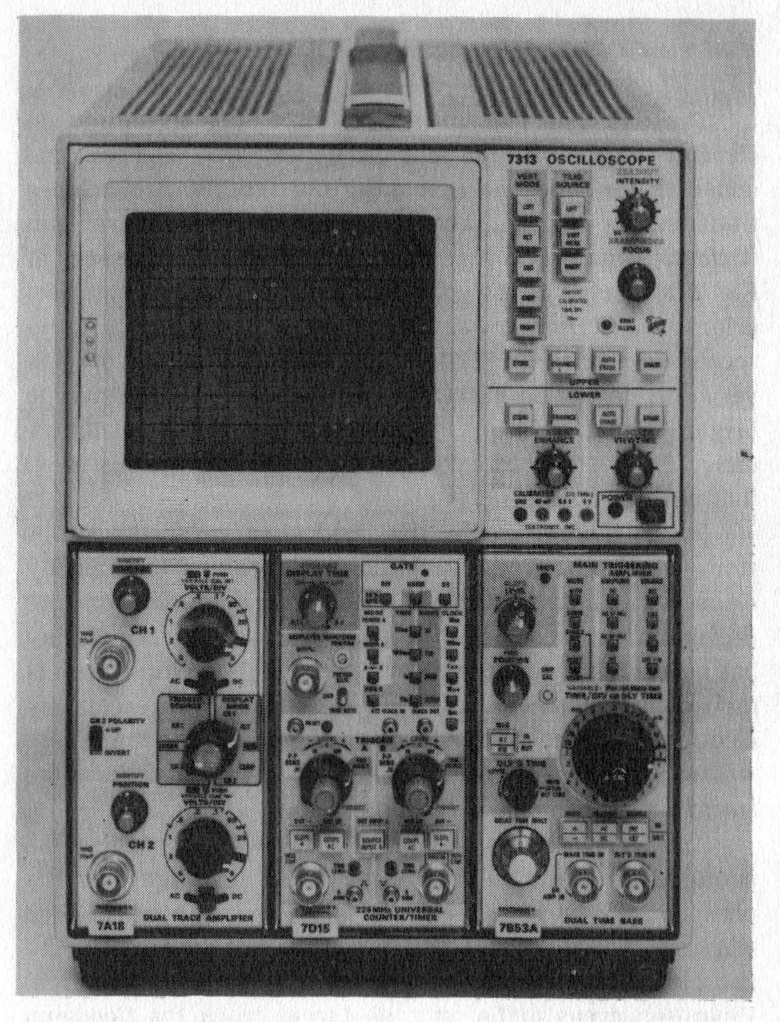

Fig. 8-27. A bistable, split-screen storage scope—the Tektronix 7313. (Courtesy Tektronix, Inc.)

the viewing time, that is, the time for which the display is stored, is continuously variable up to 12 sec. On its arrival, the signal to be viewed initiates a sweep. After each sweep, the stored display is retained and further sweeps are locked out for the viewing time selected by the view time control. After this interval, the display is automatically erased, and the time base is enabled for the next sweep. The cycle will automatically repeat itself as long as a signal is applied.

Half-Tone (Variable-Persistence) CRT

The half-tone transmission storage tube is similar in structure to the bistable tube. Two mesh-type elements are added in the half-tone tube near the faceplate to achieve transmission storage. The mesh nearest the electron-gun structure is a fairly coarse structure, which, as can be seen in Fig. 8-28, serves as the collector electrode. This element accelerates electrons toward the storage target and collects secondary electrons emitted by the target. The second mesh is very fine, having about 500 lines per inch, and it is the storage target. Using thin-film techniques, a highly insulative dielectric layer is deposited on this mesh. It is on this mesh that storage occurs.

In the storage mode, the flood guns cover the entire storage target area with a continuous stream of low-velocity electrons, as in the bistable crt. However, these electrons are prevented from reaching the phosphor screen unless a display has been written on the storage target. As the writing beam is moved across the storage target, it dislodges secondary electrons from the dielectric, as depicted in Fig. 8-28A. These written areas charge positive, while the unwritten areas of the storage mesh remain negative.

An accelerating potential of about 7 kV exists between the storage target and an aluminized layer deposited over the phosphor. Fig. 8-28B shows how this potential attracts flood electrons through the written area of the storage target to the phosphor. Thus, the phosphor glows behind the written area. Unwritten areas of the storage target block the flood-gun electrons so that these areas remain dark. There is no background glow such as we said exists in a bistable storage crt, so the half-tone tube provides a high-contrast display compared to the bistable tube.

The density of the writing beam striking the storage target determines the amount of positive charge on the dielectric of the target. The charge in turn determines the amount of flood electrons reaching the phosphor, and thus the brightness of the stored display. It is this ability to store and display variations in intensity that gives the half-tone tube its name.

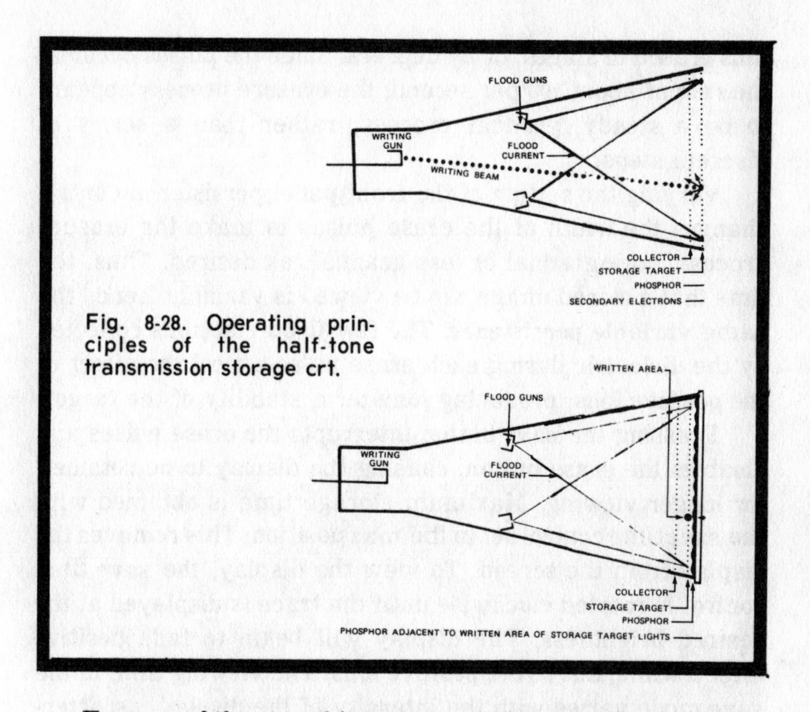

Fig. 8-28. Operating principles of the half-tone transmission storage crt.

Erasure of the stored image on a half-tone storage tube is accomplished by applying a positive pulse of about 10V to the storage target mesh. The dielectric deposited on this mesh goes positive also by capacitive coupling. However, the dielectric immediately starts to discharge back to about zero volts due to the flood-gun electrons striking it. After about a half-second, the positive pulse ends and the storage target drops back near zero volts. Again by capacitive action, the dielectric potential changes by 10V in the negative direction, and ends up at about minus 10V. The storage target is then in a ready-to-write state.

A characteristic of half-tone storage is that unwritten areas of the target begin to fade positive due to the generation of positive ions in the flood-electron system of the crt. Thus, the entire screen reaches a stored condition after a few minutes and the desired image is no longer visible. To prevent this happening and to provide optimum viewing of the stored image, the entire screen is slowly erased during operation. A series of brief erase pulses is applied to the storage target, each pulse causing a certain amount of erasure. The screen is thus erased in stages, or by degrees. Since the pulses occur at the rate of about 100 per second, the erasure process appears to be a steady, gradual process, rather than a series of discrete steps.

Varying the setting of the front panel persistence control changes the width of the erase pulses to make the erasure process more gradual or less gradual, as desired. Thus, the time that a stored image can be viewed is variable, hence the name **variable persistence**. The few flood electrons collected by the dielectric during each erase pulse cancel the effect of the positive ions, producing long-term stability of the target.

Pressing the **save** button interrupts the erase pulses and disables the erase button, causing the display to be retained for longer viewing. Maximum storage time is obtained with the **save** time control set to the **max** position. This removes the display from the screen. To view the display, the **save** time control is rotated clockwise until the trace is displayed at the desired brightness. The display will begin to fade positive after a while, due to the positive ions. The viewing time in the **save** mode varies with the intensity of the display, as determined by the **save** time control.

SAMPLING SCOPES

We saw earlier how the response of a conventional scope to fast-rise pulses and high-frequency signals is limited by the bandwidth of the vertical amplifier. This bandwidth is itself limited by the frequency limitations of the transistors used in the amplifier. In the present state of the electronics art, it is possible to build vertical amplifiers with bandwidths of up to 500 MHz and rise times as short as approximately 0.8 nsec. If sinusoids with higher frequencies or pulses with faster rise times are to be viewed, **sampling techniques** must be employed. The techniques we are about to describe permit viewing signals whose frequency or rise time exceeds the capability of the crt itself! Sampling techniques are used to display repetitive events, that is, events that recur at regular intervals; and can permit the display of recurring sinusoids having frequencies up to approximately 14 GHz and pulses having rise times down to approximately 30 psec (picoseconds).

A sampling scope is analogous to a stroboscope, an instrument that permits visual observation of rapidly rotating machinery by momentarily lighting it at slightly advanced positions on successive revolutions.

A sampling system looks at a small portion of a waveform, remembers the amplitude for as long as desired, and presents a display of the instantaneous amplitude, all without amplifying the signal directly. It looks at the waveform again slightly later in time, presents a new portion of the display, and ultimately reconstructs the entire waveform by continuing in this manner.

Fig. 8-29 shows the reconstruction of a repetitive square wave. Note that the crt display consists of a series of dots rather than the continuous trace characteristic of a conventional crt display. In the figure, a series of brief sampling pulses is superimposed on the input waveform. The pulse samples are displayed on the crt, because at the peak of each sampling pulse the crt is unblanked and a spot appears on the screen. A large number of such dots, typically 50 to 500, form the display, an **expanded** replica of the input signal. Notice that while the horizontal distance between two successive dots appears to represent the time between two points on a single input pulse, it actually represents the time between two sampling pulses. For this reason, a sampling scope is said to be a nonrealtime scope, while a conventional scope is a

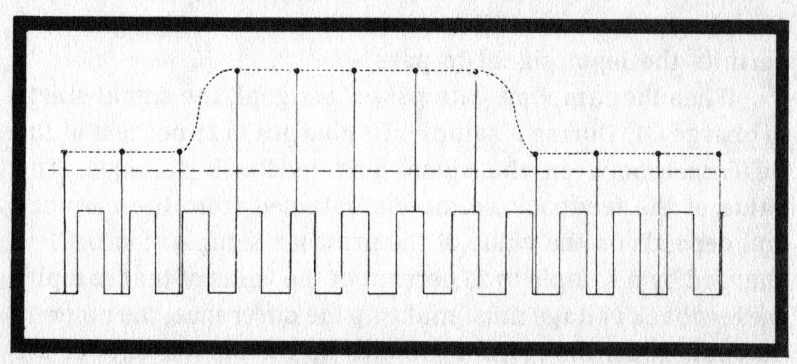

Fig. 8-29. In sampling, a series of brief sampling pulses is superimposed on the input waveform. At the peak of each sampling pulse, the crt is unblanked and a dot is displayed on the screen.

realtime instrument. Notice that it is impossible to measure the duration of a single input pulse using the sampling method illustrated in Fig. 8-29. It is, of course, possible to measure the display and obtain a statistical average of the pulse durations, which might be almost perfectly identical.

The sampling scope operates from repetitive signals, but not necessarily signals with a constant repetition rate. A small portion of each cycle of the signal is measured and a dot is displayed, which indicates the amplitude of the sampled portion of signal. The dot is horizontally positioned proportional to the point in time-space sampled.

The special circuitry of a sampling scope consists of **sample-and-hold** and **timing** circuitry. The sample-and-hold circuitry accepts the input signal, adds sampling or interrogating pulses to it as illustrated in Fig. 8-29, sends the samples along to the vertical amplifier, and holds in its memory the amplitude of the previous sample. The use of a hold memory makes it possible for a sampling system to look at the incoming signal, remember it, and then make a display correction. Using this system, it is not necessary for the output to make a transition all the way from zero for each sample, but only from the level of the previous sample.

A simplified sampling system is shown in Fig. 8-30. In this circuit, the sampling bridge-type gate is held reversed-biased except during the short sampling, or interrogation pulse duration. While the gate is reverse-biased, the signal is prevented from getting through to the first amplifier. When a sampling pulse occurs, it forward-biases the bridge gate and permits the input signal to pass.

When the sampling gate passes a signal, the signal starts to charge C1. During a sample, C1 charges to 25 percent of the **difference** between the signal and feedback voltages. The value of the feedback voltage is obtained from the memory and depends on the value of the previous sample. Since C1 is charged by a sample to 25 percent of the value of that sample, the feedback voltage must make up the difference, the other 75 percent. Thus, the memory makes up for the inability of the capacitor to charge during the short sampling time to the full value of the sample. The reason for keeping the sampling time

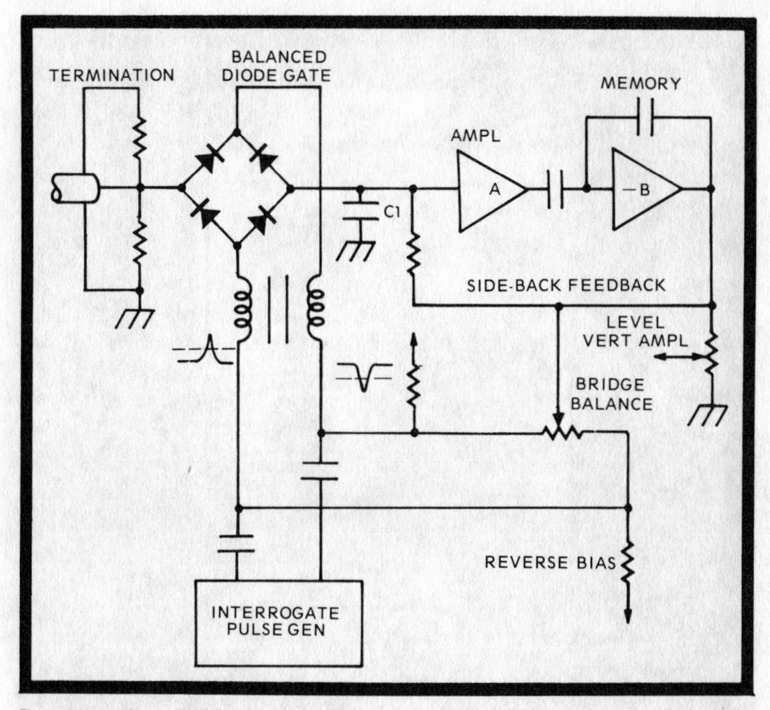

Fig. 8-30. A simplified sampling circuit using a bridge-type sampling gate. (Courtesy Tektronix, Inc.)

short is that information is lost if the sampled signal changes significantly during the sampling interval. A plug-in sampling unit for performing sample-and-hold functions is shown in Fig. 8-31. This sampling unit, the Tektronix 7S11, is used with 7000-series Tektronix scopes such as the previously mentioned 7704 and 7904.

Although the operating features of sampling scopes vary from maker to maker, and among the different models of any individual manufacturer, there are some features that are common to most. Thus, there are also some operating controls that are common to most sampling scopes. We will discuss some of these features and controls as found in the Tektronix 7S11 sampling unit of Fig. 8-31.

The **sampling head** of the 7S11 sampling unit is itself a plug-in unit, which occupies much of the lower half of the sampling module. On its front panel is the cable connector by means of which the input signal is applied to the scope. A

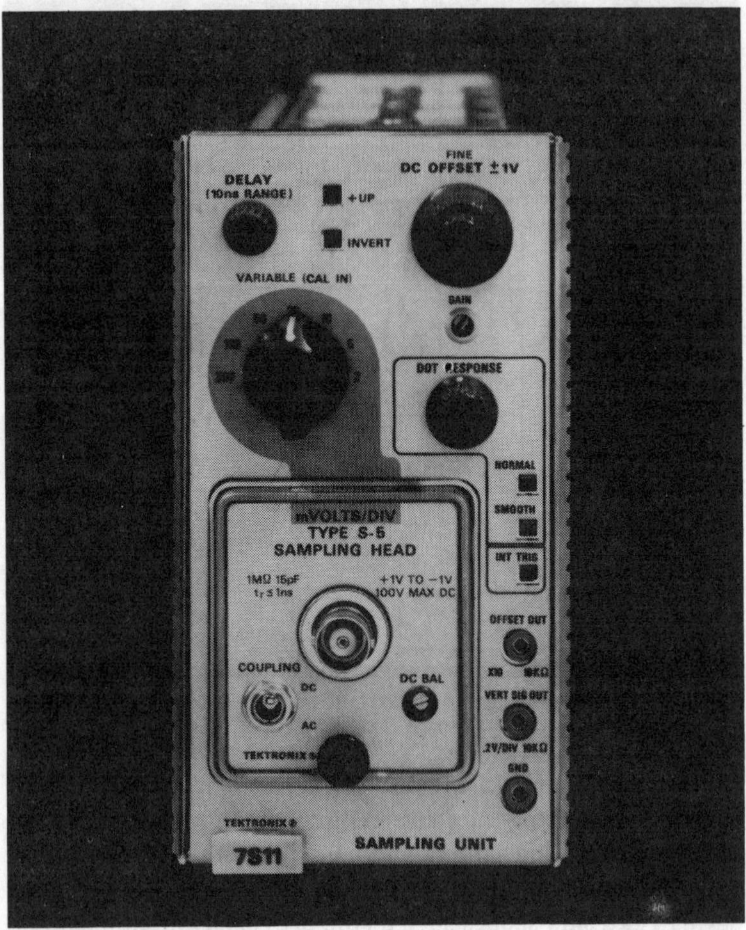

Fig. 8-31. The Tektronix 7S11 sampling unit plug-in and S-5 sampling head. (Courtesy Tektronix, Inc.)

number of sampling heads are available, covering a bandwidth from 350 MHz to 14 GHz. Sampling heads are available with several different types of input. There are 50-ohm inputs, high-impedance inputs, and active- or passive-probe inputs. The upper frequency of the signal to be viewed and the impedance of the signal source are the main considerations in the choice of a sampling head.

As in a conventional scope, the deflection factor is selected by a combination of a step attenuator and a vernier. However, the minimum sensitivity of sampling scopes is ordinarily much higher than the minimum sensitivity of conventional scopes. For conventional scopes this specification is typically a few tens of volts per centimeter; for the 7S11 sampling unit it is around 200 mV per cm. Also, the manufacturer specifies a maximum input voltage, which must never be exceeded, usually on the order of 1 or 2V. If the voltage rating is exceeded, the sampling diodes may be destroyed. Using attenuators that attach to the probe tip, larger signal voltages may be attenuated to a safe level and applied to the sampling head for measurement.

Most sampling units have a 50-ohm input impedance, and are normally used to view pulses from a 50-ohm source. However, many signals are not available at a low-impedance point, and would be altered greatly by the mismatch with a 50-ohm input impedance. Special active probes using transistors in the probe itself provide a match between the sampling unit and high-impedance signal sources. Since such probes reduce the bandwidth of the sampling system, an alternative is available in the form of a direct sampling system. In this system, the sampling diodes are located in the probe. A typical active sampling probe is the Tektronix S-3A, which has an input impedance of 100K shunted by 2.3 pF.

Ac coupling with the typical 50-ohm input requires a fairly large blocking capacitor, and this degrades the rise time. To get around this problem, the dc offset control, which adds a dc offset (bias) to the input signal is provided. As in the differential comparator discussed earlier, dc offset permits observation of a small transient or ac signal riding on a relatively large dc level by canceling the dc level. Since the dc level is canceled by the dc offset, no blocking capacitor is required. The effects of the dc offset control are greatest at high sensitivity settings.

The number of dots in the display is variable, typically over a range of 50 to 500 dots. When the dot density (number of dots per division) is reduced, transient response errors affecting the rise time of the displayed signal are introduced. These errors, often taking the form of overshoot and undershoot, can be adjusted out by means of the dot response control, which controls the gain in the sampling circuit.

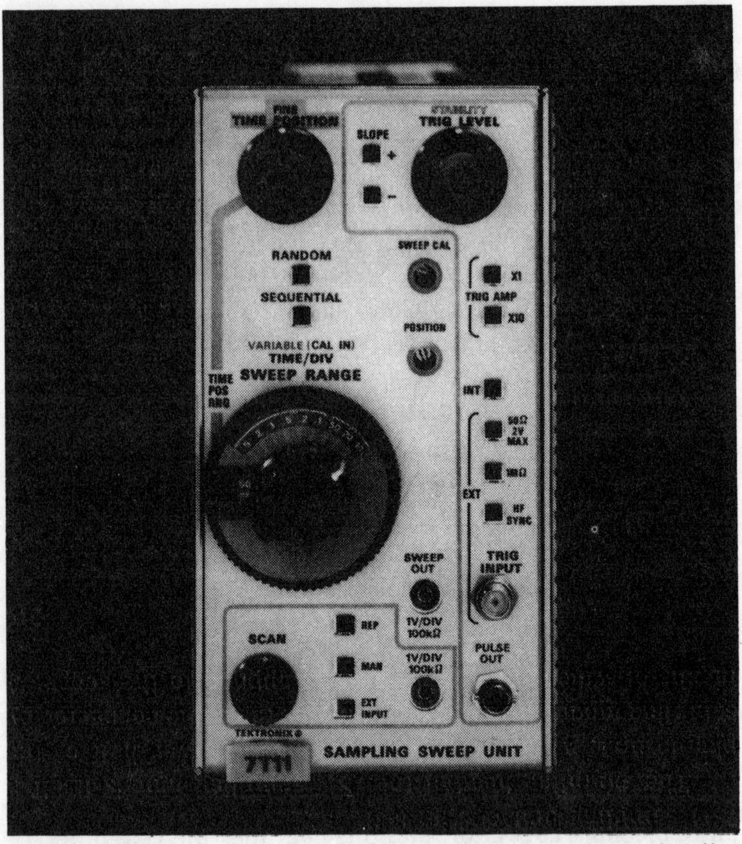

Fig. 8-32. The Tektronix 7T11 sampling sweep plug-in, used in conjunction with the 7S11 sampling unit. (Courtesy Tektronix, Inc.)

Since the sampling process entails a relatively high noise level at high gains, a provision for smoothing out noise spikes is included. The **smooth** control reduces the gain and the noise spikes by the same amount. If the gain is reduced 50 percent, the noise-spike amplitudes will be reduced 50 percent. As smoothing is increased and gain reduced, a greater sampling density is required to prevent deterioration of the displayed rise time.

So far we have discussed the sampling gate and associated holding circuitry. Now we shall consider the **timing** circuitry, which determines when the samples should be taken. In the Tektronix 7000-series scopes, the timing circuitry is contained in a separate plug-in. The Tektronix 7T11 sampling sweep unit (shown in Fig. 8-32) is the companion unit of the 7S11 sampling unit we have discussed.

Fig. 8-33 illustrates one means of determining when samples shall be taken. First of all, notice that the sampling pulses do occur at different points on the input waveform.

In the sampling sweep unit, a fast ramp is generated whose start corresponds to trigger pulses initiated by the input waveform. In some cases, trigger pulses are obtained from some external source. A staircase waveform is also generated in sampling sweep unit, and the amplitude of this signal is increased one step at the end of each ramp. Samples are taken at times when the slope of the ramp intersects the staircase. This causes successive samples to occur at later times on each pulse (or cycle) of the input waveform. Since the value of the staircase voltage is proportional to the time of occurrence of each sampled portion of the waveform, the staircase voltage may also be used to provide horizontal deflection. The lower the steps are, that is, the less their amplitude, the greater the sampling density (dots per division).

On the Tektronix 7T11 sweep unit, the triggering controls are in the upper right corner, and are similar in operation to the trigger controls of a conventional scope.

The time-per-division control of a sampling scope controls the effective sweep time per division. This control determines how many cycles of a signal there are in a presentation.

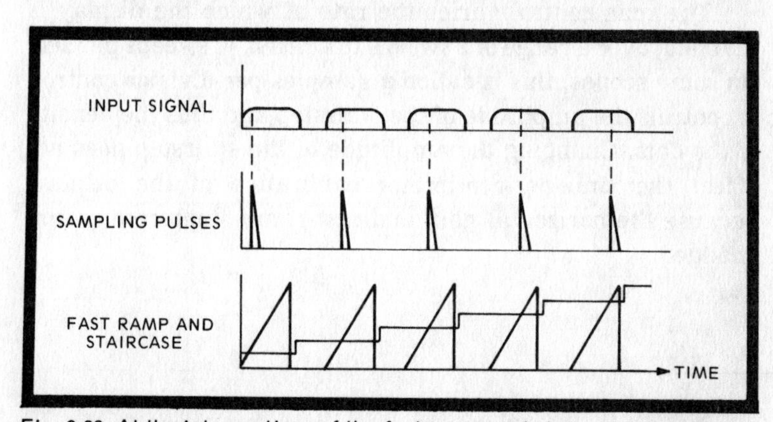

Fig. 8-33. At the intersections of the fast ramp and staircase waveforms, sampling pulses occur. The sampling pulses occur at successive points on the input signal waveform.

Because of the sampling technique, the sweep of the beam across the screen does not have to be as fast as the calibrated sweep control indicates, thus the time per division is an effective time per division. The whole idea of sampling, of course, is to reduce the performance requirements for viewing extremely fast signals. The effective sweep speed is varied by changing the slope of the fast ramps used for timing the sampling.

The time-position ring selects the time window from 50 msec to 50 nsec in seven steps. The time window is the length of the time axis, only a portion of which may be actually displayed. The time per division controls can be used to select all or a portion of the time window for display. Each range selected by the time position ring has nine times-per-division steps in 1-2-5 sequence associated with it.

The displayed portion of the time window—first part, middle part, last part, and so on—is selectable with the time position control. This control adjusts the delay between the trigger and the start of the timing ramp, and thus provides a kind of delayed-sweep capability, as just explained.

Provision is made in the 7T11 for external triggering from low-impedance, high-impedance, and high-frequency sources. The timing unit is made compatible with any of these sources by pressing the appropriate button above the trigger input connector.

The scan control varies the rate at which the display is scanned, over a range of 2 sweeps to at least 40 sweeps per sec. On some scopes, this is called a **samples-per-division** control. It controls the amplitude of the stairstep and thus the density of the dots. Changing the amplitude of the stairstep does not affect the **time-per-centimeter** calibration of the display, because the horizontal gain and fast-ramp slope remain unchanged.

Special-Purpose Instruments

Besides being used in scopes, the crt is used as a display device in many other types of electronic instruments. In this chapter, we shall discuss the following instruments, all of which are based on the crt: **spectrum analyzers, engine analyzers, semiconductor curve tracers, vectorscopes,** and **TV waveform monitors**. These instruments are really special-purpose scopes.

SPECTRUM ANALYZERS

Just as a scope displays signal amplitude across a horizontal **time** axis, a spectrum analyzer displays signal amplitude across a horizontal **frequency** axis. Fig. 9-1 compares the time-domain display of the scope and the frequency-domain display of the spectrum analyzer. A spectrum analyzer is basically a combination of a sensitive radio receiver and the horizontal sweep section of a scope.

Spectrum analyzers are used to perform a variety of rf measurements, permitting evaluation of relative amplitudes and frequencies of the discrete components of rf signals; and providing information on bandwidths, modulation characteristics, spurious emissions, and other parameters that would otherwise be difficult or impractical to measure. Because of their complex composition spectrum analyzers are especially useful for analyzing **pulse waveforms**. In microwave systems, waveform measurement by spectrum analysis is especially important, because microwave equipment, such as radar, is frequently pulse modulated. Then too, the increasing use of pulse techniques for telemetry and digital computer equipment has increased the applications of spectrum analysis.

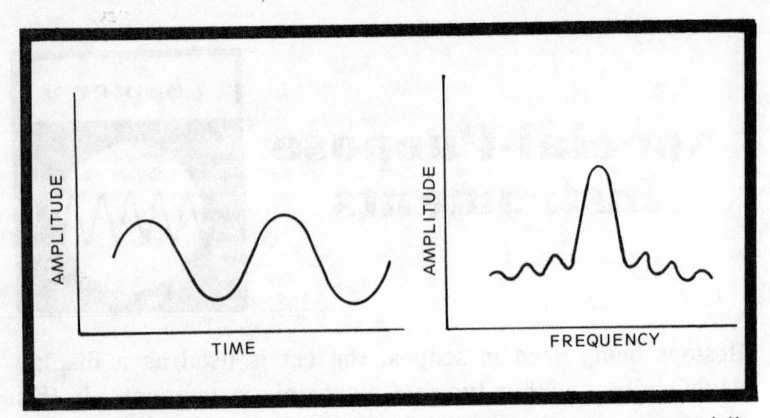

Fig. 9-1. Comparison of the time-domain display of the scope and the frequency-domain display of the spectrum analyzer.

A time-domain (or time-axis) display such as that obtained with an ordinary scope gives information on the amplitude, duration, period, rise time, and fall time of a signal. If the applied signal consists of a single frequency, information about that signal can be obtained from a time-domain display. However, if the signal consists of more than one frequency, each of which might have a different amplitude, the various frequencies will be combined into a single waveform, and the individual characteristics of each frequency will be difficult or impossible to measure.

When modulated, any fundamental frequency will produce the fundamental plus sideband frequencies; collectively, this is called a **spectrum**. The distribution of the power on these frequencies is dependent on the modulation. Ordinarily, modulation is plotted on an amplitude-and-time basis, that is, in the time domain, as shown in Figs. 9-2B, C, and D.

Let us assume that F is the fundamental frequency of an oscillator. Fig. 9-2A shows the amplitude of the fundamental plotted against time. The number of periods occurring in 1 sec determines the frequency of the oscillation. The peak-to-peak amplitude is proportional to the distance in the display between the negative and the positive peak of one cycle.

In the frequency-domain display on the screen of a spectrum analyzer, this same frequency and amplitude would be represented as a vertical pip, F in Fig. 9-2E. Points along

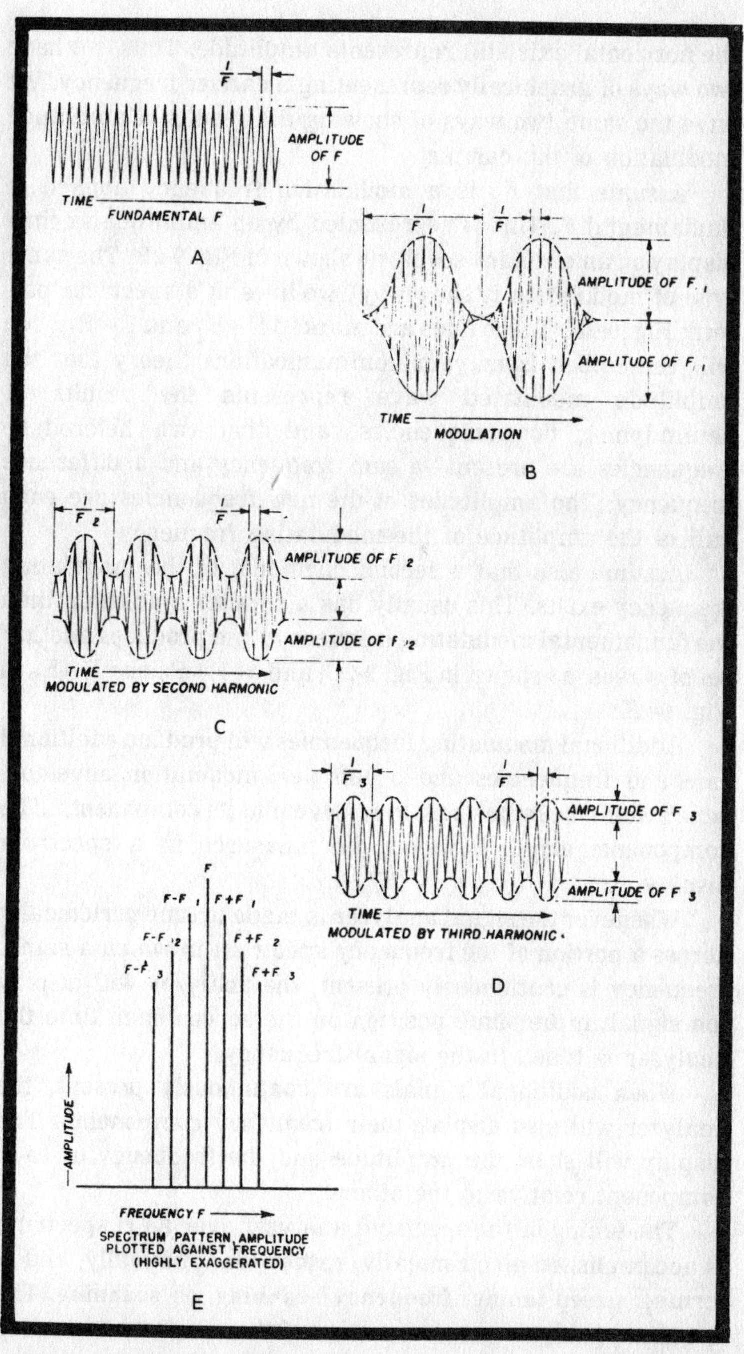

Fig. 9-2. Comparison of amplitude vs time to amplitude vs frequency.

the horizontal axis still represents amplitude. Thus, we have two ways of graphically representing a carrier frequency. We have the same two ways of showing the results of amplitude modulation of the carrier.

Assume that F_1 is a modulation frequency applied to fundamental F. This is represented by an amplitude vs time display on an ordinary scope, as shown in Fig. 9-2B. The same type of modulation is shown by **two** lines in a spectrum pattern, Fig. 9-2E. These lines are marked $F+F_1$ and $F-F_1$. You will remember from your communications theory that the amplitude modulated wave represents the results of heterodyning two frequencies, and that two heterodyne frequencies are present—a sum frequency and a difference frequency. The amplitudes of the new frequencies are each half of the amplitude of the modulating frequency.

Assume also that a second harmonic of the modulating frequency exists. This usually has a smaller amplitude than the fundamental modulating frequency; and produces another set of waves, as shown in Fig. 9-2C, and at $F+F_2$ and $F-F_2$ in Fig. 9-2E.

Additional modulating frequencies will produce additional sideband frequencies and a complex modulation envelope, which would be impossible to resolve into its components. The components could, however, be measured in a spectrum display.

Whenever the signal analyzer is made to tune periodically across a portion of the frequency spectrum in which a signal frequency is continuously present, the analyzer will display the signal in the same position on the screen each time the analyzer is tuned to the signal frequency.

When additional signals are continuously present, the analyzer will also display their frequency components. The display will show the amplitude and the frequency of each component relative to the others.

The tuning of the spectrum analyzer over an rf spectrum is accomplished electronically, rather than manually, and is termed **sweep tuning, frequency scanning,** or **scanning.** The spectrum analyzer scans a portion of the rf spectrum in step with a horizontal deflection voltage and displays on a crt any signals present, in terms of their component frequencies and their amplitudes.

Fig. 9-3 is a block diagram of a spectrum analyzer. Signal frequencies are applied to the wideband input of the mixer, where they are mixed with the swept local oscillator (LO) frequency. The mixer output at any time contains many frequency products, most of which are filtered out by the narrowband i-f amplifier circuitry. Of main interest are the first-order sum and difference frequencies. The net change in swept-oscillator frequency from the beginning to the end of the scan determines the scan width. The resolution bandwidth is the bandwidth of the i-f amplifier, and is analogous to the width of a searchlight beam or the selectivity of a receiver.

The i-f amplifier provides amplification, in addition to resolution, and passes the display signal along to the detector, which rectifies the output. The envelope of the display signal appears at the output of the detector.

Amplification of the detector output is provided by the analyzer's vertical amplifier for proper vertical deflection of the displayed signal.

The horizontal sweep sawtooth generator provides a means of electronically tuning the swept LO as well as a means of sweeping the cathode-ray beam horizontally across the screen.

In recent years, the use of a **varactor** (a voltage-controlled capacitor) and transistorized circuitry has led to the development of spectrum analyzers having but a fraction of the volume and weight of older analyzers. These newer

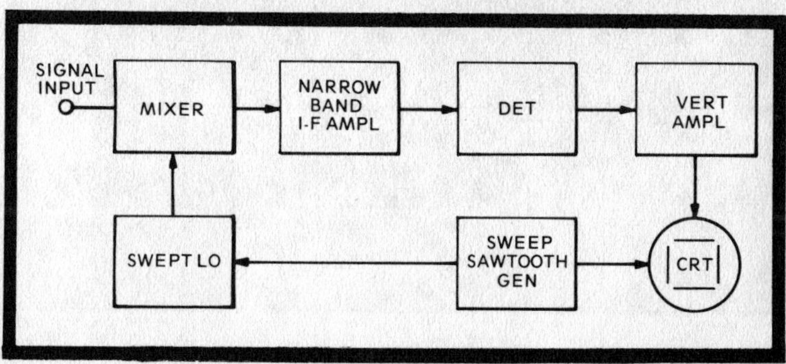

Fig. 9-3. The block diagram of a spectrum analyzer.

analyzers feature calibrated frequency span (dispersion), which permits selection of the range of frequencies to be represented by a horizontal division, accurate to 10 percent or better.

In Fig. 9-4 we have the Tektronix 1401A spectrum analyzer module, which is used in conjunction with a conventional scope. The module, which has measurement capabilities in the 1 MHz to 500 MHz range, is shown with the Tektronix 323 portable scope. These two units comprise a complete spectrum analyzer system weighing about 15 pounds.

ENGINE ANALYZERS

The most common use of the scope in engine analysis is for displaying the operation of the ignition system. Scopes used for this purpose are similar to conventional scopes, but often the controls are called by different names. Instead of a sweep frequency or sweep rate control, the engine analyzer may have a **stability** control. There may be an **expand** control in-

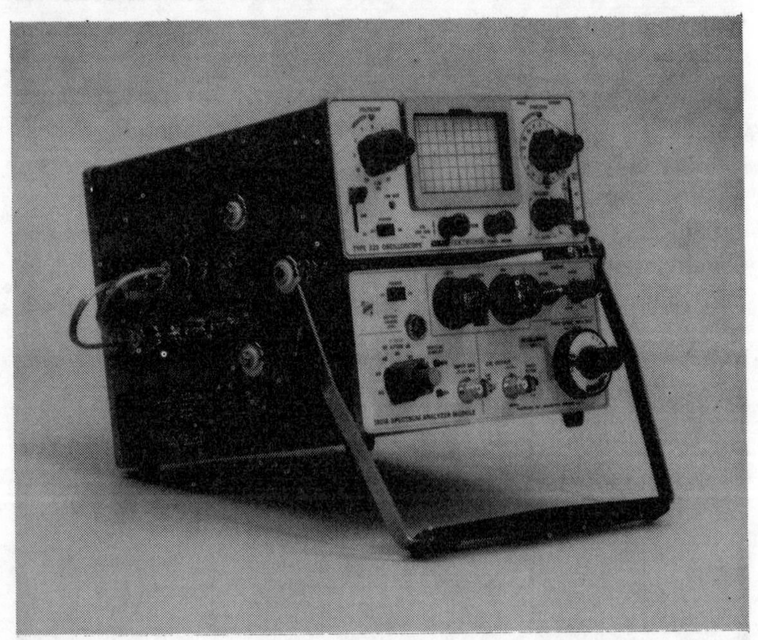

Fig. 9-4. A modern spectrum analyzer system, consisting of a portable scope and a spectrum analyzer module. (Courtesy Tektronix, Inc.)

stead of a horizontal gain control, and a **parade** control instead of a horizontal position control. Instead of a vertical gain control, there may be a **height** control. The names of the controls on an engine analyzer of this type are actually more descriptive than names of conventional scope controls, and are practically self-explanatory.

There is a surprising amount of information about engine operation that may be obtained with an engine analyzer. In fact, the information to be gained is limited mainly by the ability of the operator to interpret the meaning of observed waveform deviations from the normal wave patterns.

Ignition Displays

The analyzer is connected to the engine by connecting one lead to the main coil wire and the other lead to the number 1 spark plug wire. Usually, both leads are shielded and must be connected to ground. The next step in obtaining an ignition display is to turn on the engine analyzer. Then start the engine and adjust its speed to 1000 rpm. Use a primary-actuated tachometer (tach) for this adjustment, and remove the tach leads when the adjustment is completed. If not disconnected, the tach may cause abnormal scope patterns.

To develop a normal secondary pattern, adjust the expand and parade controls so the pattern occupies practically the entire width of the screen and is centered, as shown in Fig. 9-5. Adjust the height control so that the highest peak is at the 50 percent line, as shown. Adjust the stability control to display the proper number of cylinders and to stabilize the pattern.

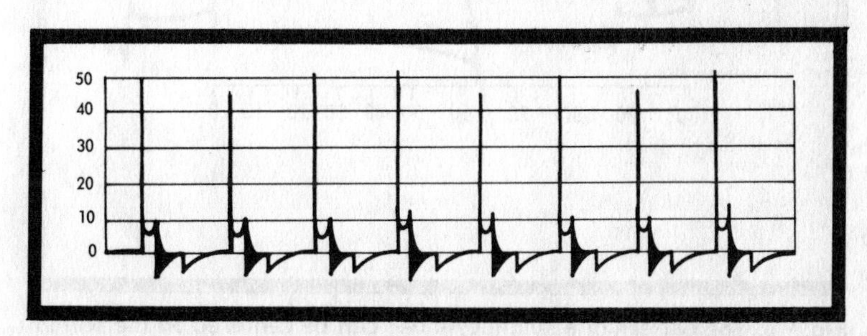

Fig. 9-5. The normal secondary pattern of an 8-cylinder engine.

Using the expand and parade controls, the pattern of one cylinder can be centered on the bottom scale of the screen. The dwell time, which is the time during which the points are closed, can be measured on the scale (Fig. 9-6), and should be between 60 and 70 percent for all engines regardless of the number of cylinders. On some engine analyzers, the scale may be calibrated in terms of dwell angle. In any case, very accurate dwell measurements are possible with this type of scope.

An open circuit anywhere between the distributor cap and a spark plug can also be detected. First adjust the expand control to crowd the cylinder patterns together at the left side of the screen. Next, adjust the height control so that the peaks register around the 50 percent line. Compare the heights of the peaks, and if any varies over one division, 10 percent, it indicates a gap between the distributor and the spark plug. This shows up as a higher-than-normal peak, and it may be caused by a plug wire being open or not making contact in the distributor cap.

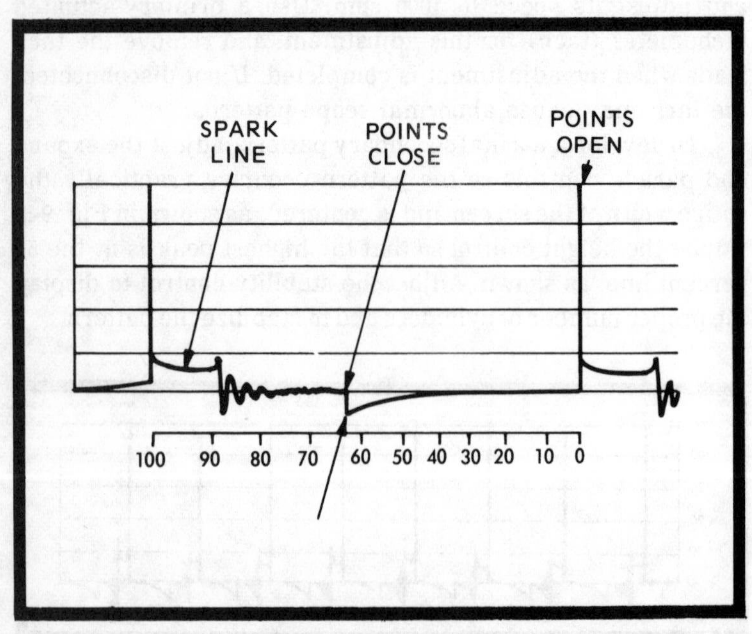

Fig. 9-6. The pattern of a single cylinder can be centered on the bottom scale, and a measurement of dwell angle made.

A test of ignition reserve is made by removing any spark plug wire except No. 1, where the analyzer is connected, and adjusting the height control until that cylinder peak reaches the top line. Fig. 9-7 shows how the voltage of the inoperative cylinder, the available secondary voltage, may be compared with the operating plug voltages. The reserve is the difference between the high peak and the other peaks, and should be 60 percent or more. Insufficient reserve may be caused by any of the following: wide plug gaps, a burned distributor rotor, burned distributor cap points, the center coil wire improperly seated, the condenser leaking, high resistance between the breaker points, or a defective ignition coil. (To test plug 1, of course, you repeat the test with the analyzer connected to another plug.)

An engine analyzer plug-in used in conjunction with the Tektronix 561B general-purpose scope of Fig. 9-8 permits the display of vibration, pressure, and crank angle, as well as ignition voltages. A simultaneous display of all four engine parameters, such as that in Fig. 9-9, may be obtained with this instrument.

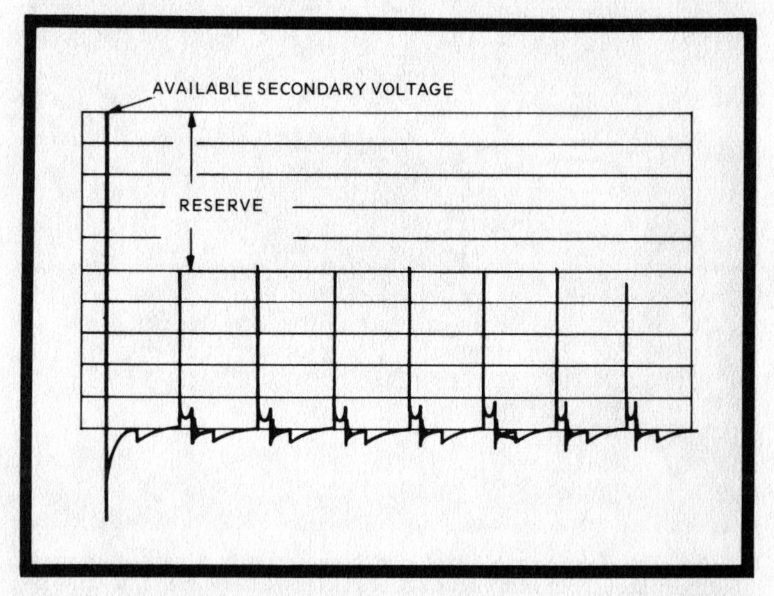

Fig. 9-7. Comparing the voltage at an inoperative cylinder with the voltage at the other cylinders gives the ignition reserve.

Ignition measurements are used for proper timing of the engine, and can detect bad wiring, bad spark plugs, point problems, defective distributor caps, defective condensers, and bad coils. Ignition measurements can also be used to determine engine rpm. Ignition measurements are made using a 1000:1 capacitive attenuator, which clamps on the secondary coil and spark plug wire. The attenuator is part of the ignition pickoff cable assembly at right center in the photo in Fig. 9-8.

Combustion Measurement and Displays

Just in front of the ignition pickoff cable assembly (Fig. 9-8) is the **rotational function generator**, the device with the large circular dial. This generator is mechanically coupled to the engine under test and generates 10, 60, and 360 deg markers. Crank-angle markers such as it generates are shown in Fig. 9-9. The rotational function generator is mechanically timed to an engine reference point by comparing the display of

Fig. 9-8. The Tektronix 561B plug-in type scope is a general-purpose laboratory scope, but with various accessories that are available, may be used in applications outside the lab. With the accessories shown, the scope may be used as an engine analyzer. (Courtesy Tektronix, Inc.)

the top-dead-center mark of the magnetic pickup from the flywheel with the 0-deg pulse generated by the function generator. This unit also generates a sawtooth ramp for displays related to crank angle, and also a waveform that is equivalent to piston volume. A pressure vs volume display such as that in Fig. 9-10 is used to determine actual engine horsepower and to detect overall problems. The area within the loop is the **mean effective pressure**, and is used to determine the horsepower.

Just to the left of the rotational function generator and slightly behind it in Fig. 9-8 is a pressure transducer, a piezoelectric type that screws into the engine block in place of a spark plug. The use of the pressure transducer in combination with the rotational function generator, permits a pressure vs volume display and a calculation of horsepower. Also, pressure measurements detect peak firing pressures, compression, early and late cylinder firing, and preignition.

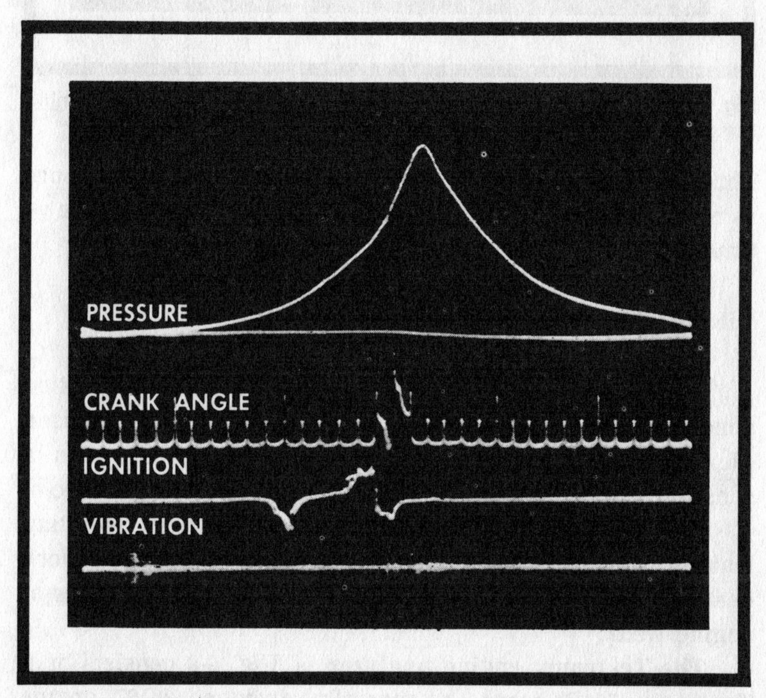

Fig. 9-9. A simultaneous display of four engine parameters on a cathode-ray tube.

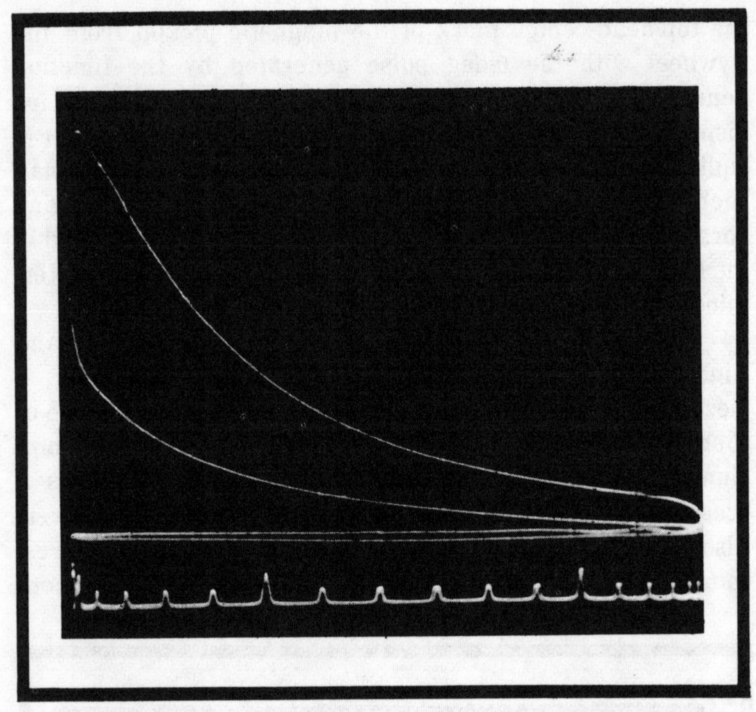

Fig. 9-10. A pressure vs volume display, useful for determining horsepower.

Three displays of cylinder pressure are obtainable: pressure vs cylinder volume, pressure vs crank angle, and pressure vs time.

Vibration Display (Engine Wear Troubles)

Immediately to the left of the rotational function generator in the front row of accessories in Fig. 9-8 is a **piezoelectric vibration generator**. This transducer has a sensitivity of 6 mV/g and a maximum acceleration spec of 1000g. Vibration measurements are useful for detecting leaking valves, knocking, cylinder wear, blowby, worn bearings, broken rings, valve flutter, and many other signs of engine wear.

The Tektronix engine analyzer of Fig. 9-8 consists of a Tektronix 561B scope, a specially designed 2B67 engine analyzer time base with a rotational function generator input, and a 3A74 engine analyzer amplifier featuring four channels, with inputs for pressure, ignition, vibration, and crank-angle markers.

VECTORSCOPE

A vectorscope is a scope that displays a flower-like waveform consisting of 10 "petals," which represents the chroma signal being fed to the guns of the crt of a color TV receiver. The vectorscope either contains or is used in conjunction with a color-bar generator, an instrument that provides a standard test signal for a TV receiver and produces on the TV screen a display consisting of 10 colored bars, as illustrated in Fig. 9-11.

The ideal test signal of the color-bar generator, which, when applied to the antenna terminals of the TV results in the picture of Fig. 9-11, produces waveforms such as those in Fig. 9-12 at the receiver's color demodulator output. Note that these waveforms are 90 deg out of phase.

Fig. 9-13 shows that when two sine waves 90 deg out of phase, but of the same frequency and amplitude, are applied to the deflection plates of a crt (one signal to the horizontal plates, one to the vertical plates), the Lissajous figure produced on the crt screen is a circle. Applying the R minus Y, and B minus Y signals from the chroma demodulator output of

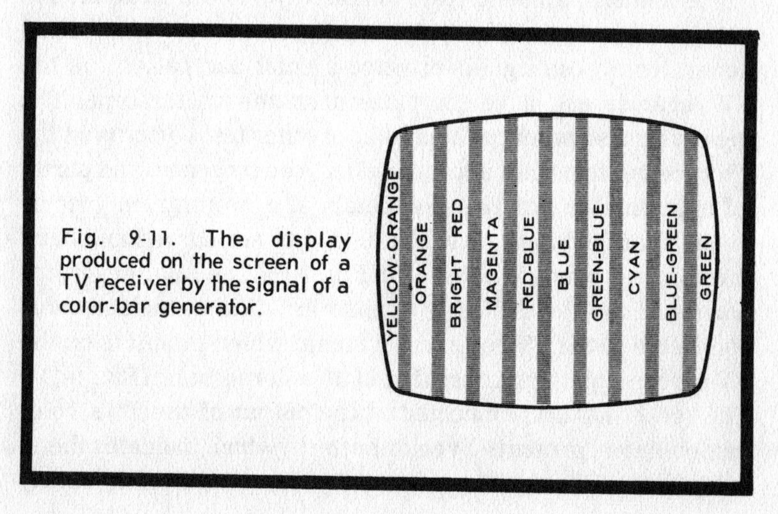

Fig. 9-11. The display produced on the screen of a TV receiver by the signal of a color-bar generator.

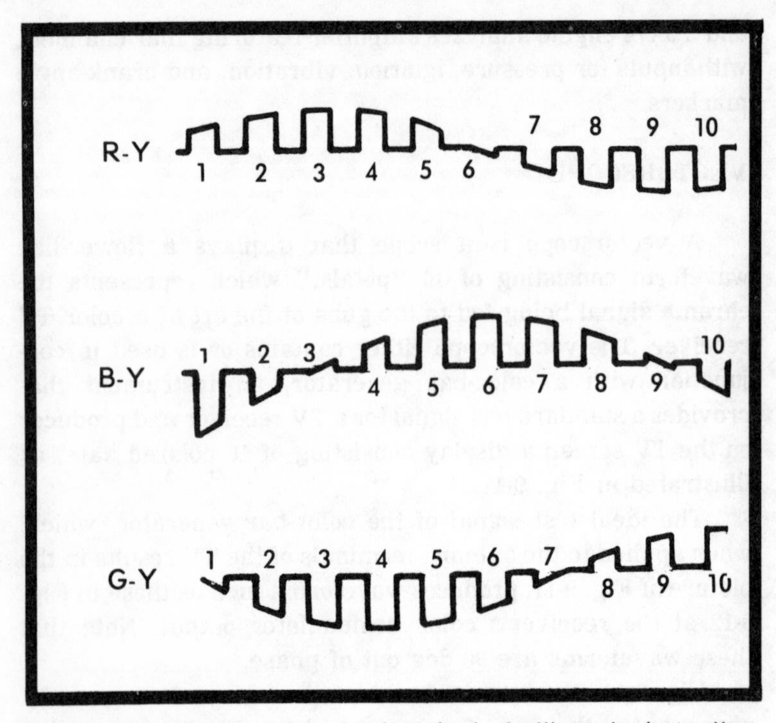

Fig. 9-12. Ideal signals at the color demod output with a color-bar pattern input to the TV receiver.

a TV to the vertical and the horizontal deflection plates of a vectorscope crt will produce the Lissajous pattern of Fig. 9-14.

A modern, all-solid-state vectorscope is the Heathkit IO-101 in Fig. 9-15. This instrument contains a test signal generator, whose signals produce a color-bar pattern on the TV receiver and a vector pattern on the vectorscope. The generator also produces a number of other test patterns on the TV receiver to permit accurate color, convergence, and purity adjustments. In practice, the petals of a vectorgram (vector display) are not ideal as in Fig. 9-14, but are curved and have more or less pointed peaks, as in Fig. 9-15. The color generator portion of the Heath IO-101 "transmits" (via a shielded cable connected to the TV receiver) a signal which produces on the TV screen a pattern consisting of 10 colored bars (Fig. 9-11). The scope portion, connected at the output of the TV's color demodulator, presents a vector pattern, which indicates the 10 color bars. The signal produced by the generator is nearly

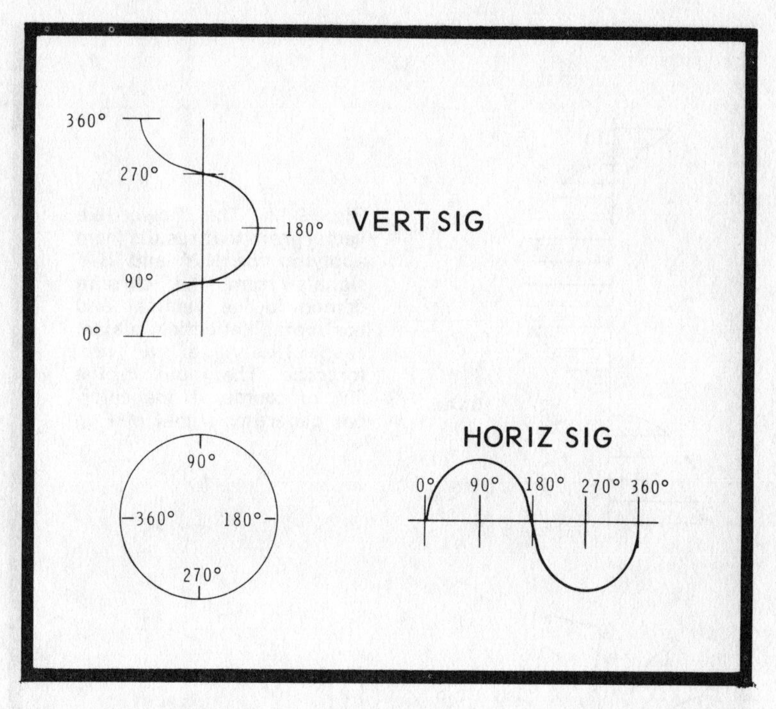

Fig. 9-13. Two identical sine waves 90 deg out of phase produce a circular Lissajous figure.

ideal, and any variation in the vector pattern from the ideal pattern of Fig. 9-14 may be assumed to have been produced in the TV receiver.

Even for a normally operating receiver the vector patterns are not ideal, since a receiver does not have to operate to perfection to produce a normal picture. Vectorgram waveforms are different in different receivers and are due to normal circuit tolerances. We are not concerned with normal circuit tolerances, but if a vectorgram shows distortions that are out of normal tolerance, then we know that a circuit defect is present, and often have a clue as to where it may be found.

As can be seen in Fig. 9-15, the vectorscope screen is marked with ten numbered angle lines. These are spaced 30 deg apart to aid in measuring the demodulation angle. The reference signal (burst) is at 0 deg. The third petal is the R minus Y signal and is 90 deg away from the burst signal. The sixth petal is the B minus Y signal, and it is ideally 180 deg

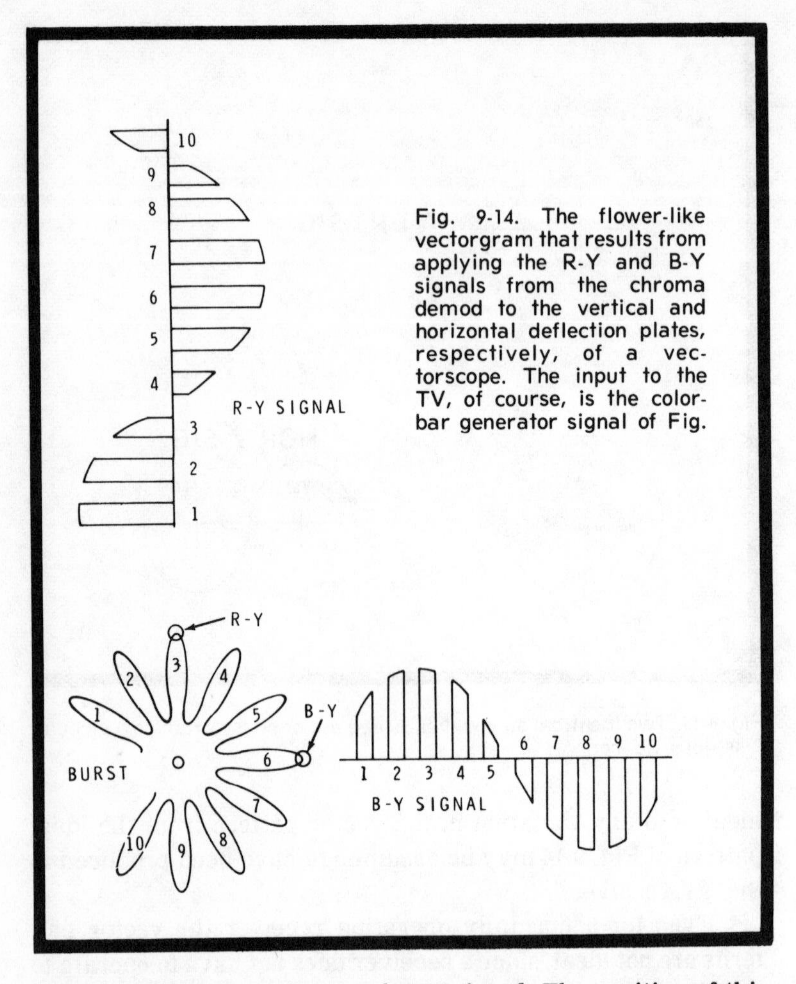

Fig. 9-14. The flower-like vectorgram that results from applying the R-Y and B-Y signals from the chroma demod to the vertical and horizontal deflection plates, respectively, of a vectorscope. The input to the TV, of course, is the color-bar generator signal of Fig.

removed from the reference burst signal. The position of this petal varies between about 80 and 120 deg, depending on the design of the TV demod. You should check the service data for the TV for the proper angle before you assume there's a circuit defect if B minus Y falls outside the limits. The tenth petal is the G minus Y signal.

As can be seen in Fig. 9-15, the vector pattern has an inner circle, ideally. In practice, the inner pattern is often an ellipse, and the size of the ellipse is directly related to the TV demodulator's **bandwidth**. The wider the bandwidth, the larger the ellipse. Also, the inner pattern is a circle only when the R minus Y and B minus Y signals have had the same gain.

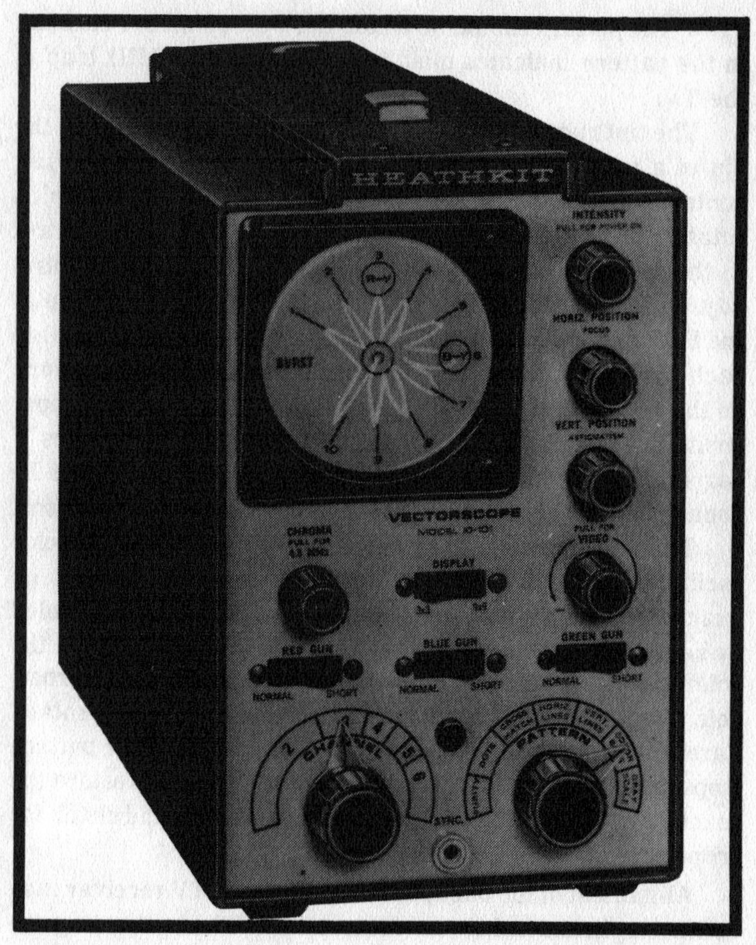

Fig. 9-15. An all-solid-state vectorscope. (Courtesy Heath Co.)

There are a number of important things to look for in a good vectorgram:

1. The sides of the petals should be straight, extending from the center of the vectorgram to the end of the petal. The petals should be neither bent or bowed, nor should they overlap each other.

2. The ends, or tips, of the petals should be flat **and brighter** than the sides of the petals.

3. The inner circle should be small, with a dot in the center. The sides of the petals should converge at the center circle.

4. The pattern should be clean and well-defined. Fuzziness in the pattern indicates misalignment of the 4.5 MHz trap in the TV.

The tint range of a TV receiver may be adjusted with the aid of a vectorscope for the proper flesh tone when the tint control is set to the center of its range. As the tint control is rotated, the vector pattern is rotated also. With the tint control to the center of its range, and with the burst phase control adjusted so the third (R minus Y) petal points straight up at the R minus Y vectorscope marker, rotate the tint control to each end. The R minus Y petal should move at least one mark on the vector graticule, or 30 deg on either side of its proper position. If the tint control is to one side when the R minus Y bar is in its proper position, straight up, there will not be enough tint range to assure proper flesh tones on all stations.

The vector pattern may be used to set the 3.58 MHz color oscillator (the color subcarrier regenerator in the TV) to its exact frequency. First, the reactance stage must be disabled by removing the burst amplifier tube or transistor. Then the color killer circuit must be disabled. There will be a barber-pole effect on the TV screen, and the vectorgram will rotate. Carefully adjust the 3.58 MHz oscillator coil until the pattern stops, or is slowed to the greatest extent possible. Restore the reactance stage and the color killer to operation, and check for proper color on the receiver screen.

Alignment of the bandpass amplifier of a TV receiver may be accomplished using the vectorscope. When adjusting the bandpass transformer, the color level control should be set to approximately its midpoint, since an excess of color signal can overload the amplifier and make alignment impossible. Obtain a vector pattern and adjust the bandpass transformer for optimum vectorgram shape and amplitude. In many TVs, there will be a single bandpass transformer with two adjustment slugs. Usually, the bottom slug is adjusted for the straightest petal sides down to the center of the pattern. The top slug is normally adjusted for straightest sides and maximum amplitude of the vector pattern.

Next, adjust the chroma takeoff coil for maximum undistorted amplitude and straight sides of the petals. Observe all of the petals to make sure that you are not optimizing some of them at the expense of the others. If the vectorgram appears fuzzy, that is, if it lacks good definition, check the alignment of the 4.5 MHz trap. Misalignment of the trap can cause vectorgram fuzziness and sound interference in the color when the receiver is tuned to a station.

Besides receiver adjustments, the vectorscope can be used for troubleshooting. Defects in color circuits of a color TV will show up in the vector pattern. If there is a weakness or loss of red, this may be evidenced on the vectorscope by the vertical amplitude of the pattern being low. A check of the R minus Y difference amplifier and its demodulator circuits should locate the trouble. A weak blue (B minus Y) signal will

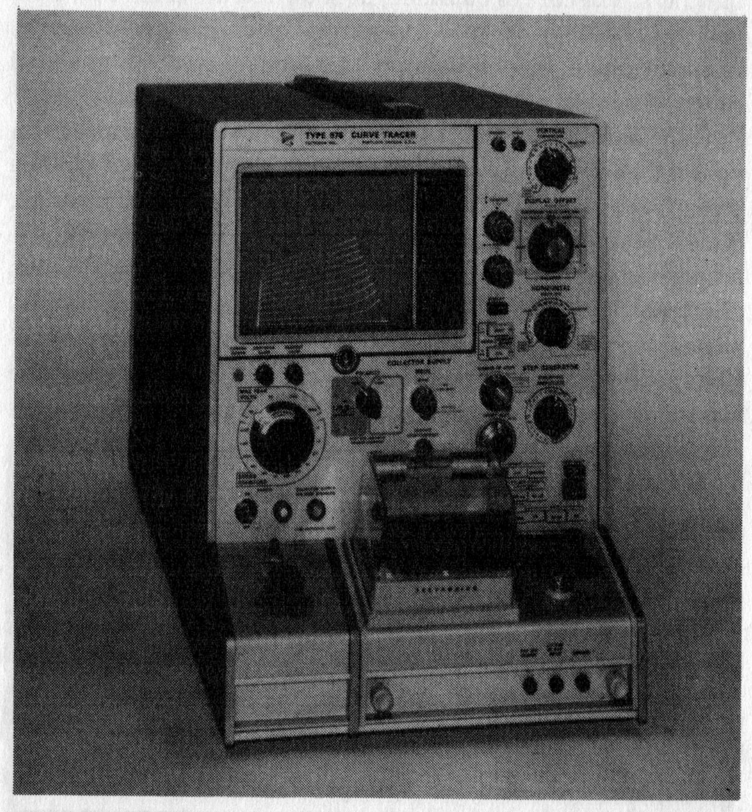

Fig. 9-16. A semiconductor curve tracer, used to display the characteristic curves of semiconductor diodes, transistors, and FETs. (Courtesy Tektronix, Inc.)

show up as a vector pattern that lacks proper width. For this, check of the B minus Y difference amplifier and its demodulator.

A vectorscope is one of the most useful of all instruments for testing color receivers. Its main advantage is that it shows the overall operation of two chroma channels simultaneously, and saves time.

SEMICONDUCTOR CURVE TRACER

A semiconductor curve tracer, such as the Tektronix 576 in Fig. 9-16, is a special-purpose scope used to display the characteristic curves of semiconductor diodes, transistors, and FET. The curves obtained, as shown on the crt in Fig. 9-16, are very similar to those found in spec sheets. Voltage is graphed along the horizontal axis and current is graphed along the vertical axis.

The basic concept of using the scope to display voltage vs current (E-I) characteristics is illustrated in Fig. 9-17. The sweep source provides a full-wave rectified sine-wave voltage, which is applied between the collector and emitter of the transistor under test. This signal is also applied to the X-axis (horizontal) amplifier of the scope, and provides horizontal deflection of the cathode-ray beam. Thus, the horizontal-deflection voltage is proportional to the collector voltage of the transistor. As the collector voltage rises and the beam sweeps

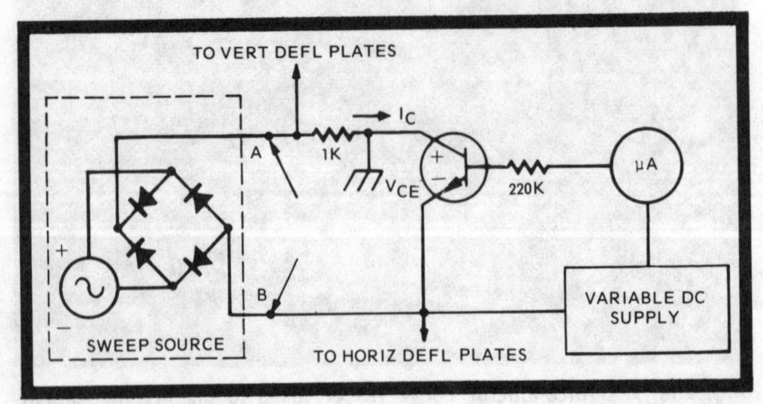

Fig. 9-17. Basic concept of using scope to display current vs voltage characteristics. (Reverse leads at points a and b to test pnp transistor.)

across the crt screen, the collector current also rises. The rising collector current causes a rising voltage drop across the 1K current-sampling resistor, which is applied to the Y-axis (vertical) amplifier. So the writing beam deflects horizontally in response to the collector voltage and vertically in response to the collector current, and a graph of current vs voltage is traced out on the crt.

Why is the signal source, or sweep source, normally sinusoidal? A sinusoidal output provides a single-frequency excitation, and, hence, a static, or dc, characteristic. With a static display, voltage and current will always be in phase; otherwise, they will be out of phase, and a closed loop rather than a single line will be displayed.

The sampling resistor not only produces a vertical signal proportional to the collector current, it limits the current and protects the device being tested. If the maximum sweep voltage is V_m, the current flowing into the collector is limited to V_m divided by R, where R is the value of the sampling resistor. The selection of V_m and R provide a boundary for the characteristic curves, which corresponds to a diagonal connecting the right hand ends of the curves in Fig. 9-16.

The full-wave bridge rectifier in Fig. 9-17 has two purposes. First, it provides the symmetrical load needed by the sine-wave generator. Without the bridge, the asymmetrical E-I characteristic of the transistor can cause difficulty in signal generators with capacitively coupled outputs. Secondly, the rectified sine wave of the bridge produces only one polarity of voltage at the collector terminal. This is desirable because the maximum safe voltages for semiconductors are usually different for opposite polarities.

In Fig. 9-17, a variable dc supply is used to set the value of base current during the sweep of the collector voltage. This results in only one curve being displayed. Using this circuit, a simultaneous display of several E-I curves such as the one in Fig. 9-16 is not possible. To produce such a graph with the circuit of Fig. 9-17, it would be necessary to make a graph of the display by hand or to make a multiple-exposure photograph, resetting the base supply for each curve.

Fig. 9-18 shows how a display of a **family** of characteristic curves such as that in Fig. 9-16 is obtained. Again, the sweep signal is a rectified 60 Hz sine wave. On the Tektronix 576 curve tracer, the polarity switch selects the polarity of the sweep signal applied to the collector. Maximum collector voltage V_m is selected by the **max peak volts** control, and the value of the sampling resistor is selected by the **series resistors** control. The sweep voltage is applied to the collector and emitter terminals of the test panel (Fig. 9-18). Thus, the driving-point characteristics of devices other than transistors may be observed by connecting such a device between the collector and emitter terminals.

To produce a family of curves rather than a single curve, the 576 tracer changes the value of the base current in synchronism with the sweep voltage. A multiple curve display is obtained in a manner similar to the way a multiple-trace display is obtained in the alternate mode of a multiple-trace scope.

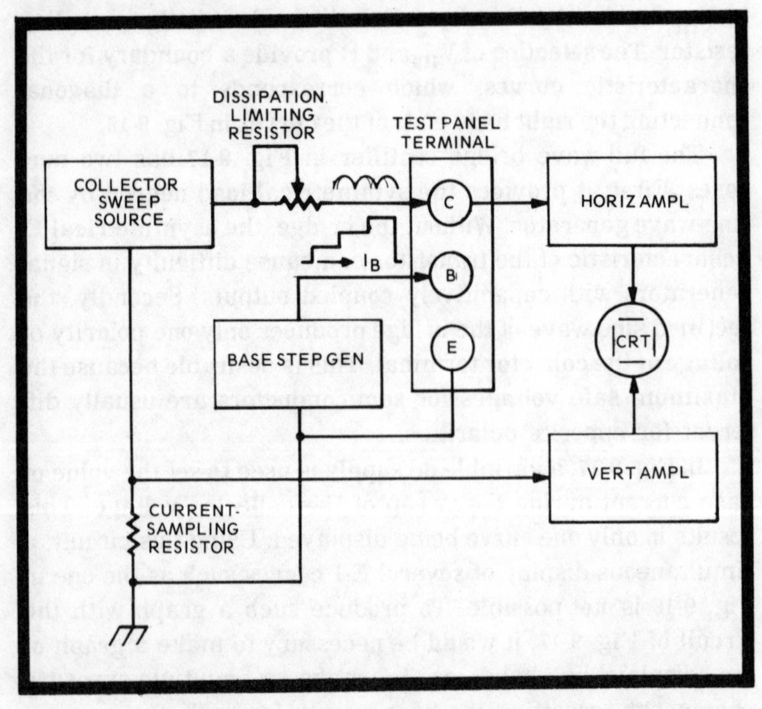

Fig. 9-18. A simplified block diagram of a curve tracer.

Fig. 9-19 illustrates how a family of characteristic curves is generated in the curve tracer. The repetition rate of the curves is determined by the 60 Hz sweep voltage. This rate is sufficiently fast to prevent flicker in the display. During each collector sweep, the **staircase current generator** provides a constant base current. The staircase current is raised one step after each sweep. In Fig. 9-19, three steps are shown. One curve of the family is produced for each step, and then the process repeats. Since the 60 Hz sine wave is full-wave rectified, there are 120 sweeps per second. The repetition rate of the family of curves is thus 120 divided by the number of curves in the display. In the situation depicted, a complete family of curves will be traced out 40 times per second, since three curves are produced. As many as 10 curves may be displayed on the Tektronix 576, or as few as 1.

Either a single family or a repetitive family may be selected. The single-family setting will display one complete

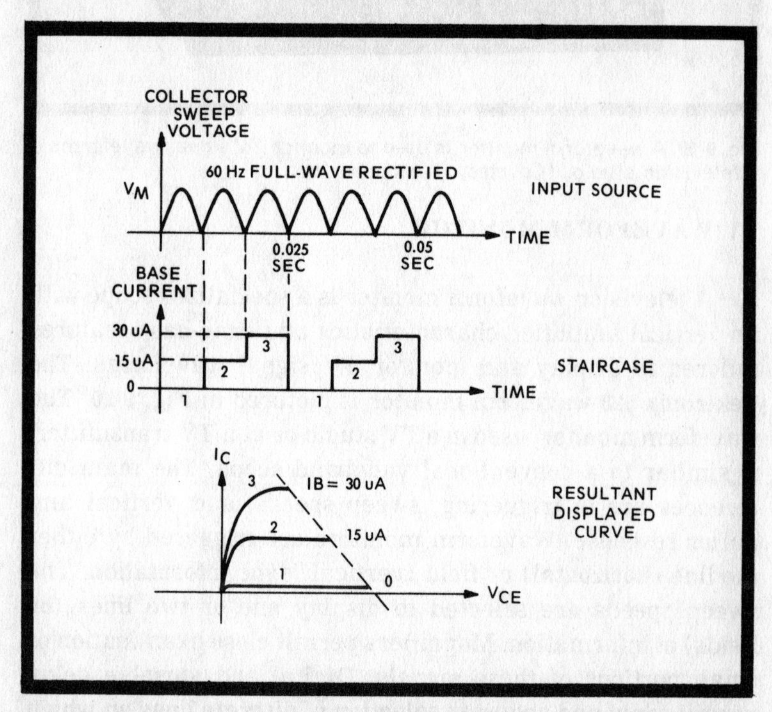

Fig. 9-19. How a family of characteristic curves is generated in a curve tracer.

set of characteristics each time a switch is pressed. This low duty cycle minimizes thermal effects on the transistor under test. The amplitude of the base-current staircase steps, the number of steps, and the polarity of the steps are adjusted by the **step generator** controls on the right side of the curve tracer, Fig. 9-16.

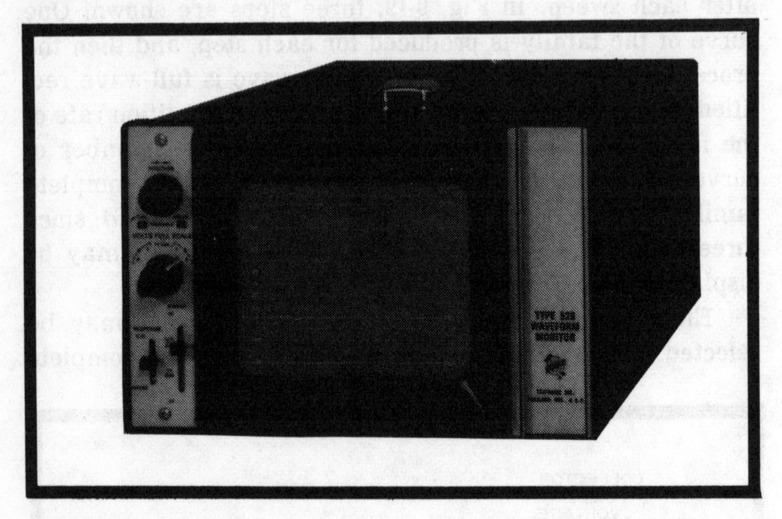

Fig. 9-20. A waveform monitor is used to monitor TV signal waveforms in a television studio. (Courtesy Tektronix, Inc.)

TV WAVEFORM MONITOR

A television waveform monitor is a specialized scope with its vertical amplifier characteristics and time-base features tailored to display and monitor TV signal waveforms. The Tektronix 528 waveform monitor is pictured in Fig. 9-20. The waveform monitor, used in a TV studio or at a TV transmitter, is similar to a conventional wideband scope. The main differences are in triggering, sweep speeds, and vertical amplifier response. Waveform monitors are triggered by either the line (horizontal) or field (vertical) sync information. The sweep speeds are selected to display one or two lines (or fields) of information. Magnifiers permit close examination of small portions of these signals. Digital and variable delay permit rapid and accurate selection of discrete lines on which any of a number of test signals may be placed.

Vertical response in a waveform monitor must be very flat within the 6 MHz video band. Usually, the response is flat within 1 percent. To achieve this, waveform monitors have a minus 3 dB-down bandwidth of around 18 MHz. Special filters may be used to further limit the response. A notch filter at the color subcarrier frequency enables viewing of the chrominance (color) information separately. A high-pass filter permits measurement of differential gain with the waveform monitor. Differential gain is the amplitude change of the color subcarrier signal component as it changes from a low (black) to a high (white) luminance level. A modulated staircase signal is used to make this measurement.

Although the highest video signal frequency is limited by noise and the need for spectrum conservation, the composite TV signal is **not** considered in the frequency domain, for measurement purposes. Video signals are basically nonsinusoidal. Thus, time domain measurements as normally made with a scope permit measurements that can be correlated with such picture impairments as smear and streaking. For example, tilt in a 10 usec square wave fed through a TV system produces picture streaking from left to right. Tilt in a 10 usec square wave produces a variation in picture shading from top to bottom. Distortions may be broken up into four time categories, each characterized by a different picture impairment:

1. **Short-time distortions** (0.125 to 1 usec) affect picture sharpness, or resolution. Undershoots on pulses in this category make the picture blurry. Overshoots, if not too great, may enhance the resolution. Ringing causes echoes or halos.
2. **Line-time distortions** (1 to 50 usec) cause horizontal streaking in the picture.
3. **Field-time distortions** (50 usec to 16 nsec) cause shading in the vertical direction.
4. **Long-time distortions** (greater than 16 ms) cause picture flicker.

Whether distortions of a particular category (as listed above) are present may be determined by passing a pulse whose duration is within the limits specified for the type of distortion through the equipment to be checked. To check for

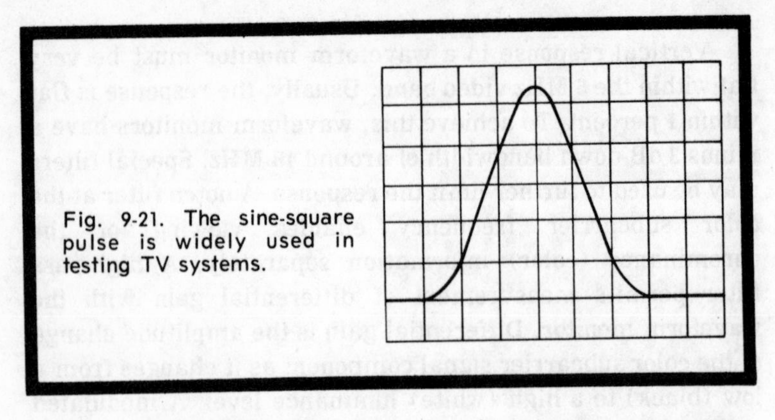
Fig. 9-21. The sine-square pulse is widely used in testing TV systems.

line-time distortion, for example, one would use a pulse whose duration was between 1 and 50 usec. Ordinary fast-rise square pulses are not used, because these pulses have a very wide band of frequency components. In television, the limit on bandwidth, 4.2 MHz, makes this type of test signal limited in usefulness. Instead of a fast-rise square wave, a sine-square pulse, such as that illustrated in Fig. 9-21, is frequently used. The energy distribution in this type of signal drops off very sharply at a certain frequency, as shown in Fig. 9-22. Also, within the passband of the signal, the energy is rather evenly distributed. This type of signal is used because it can be well confined to the passband of a television system. A sine-square pulse of the proper duration may be used to test for short-time distortion.

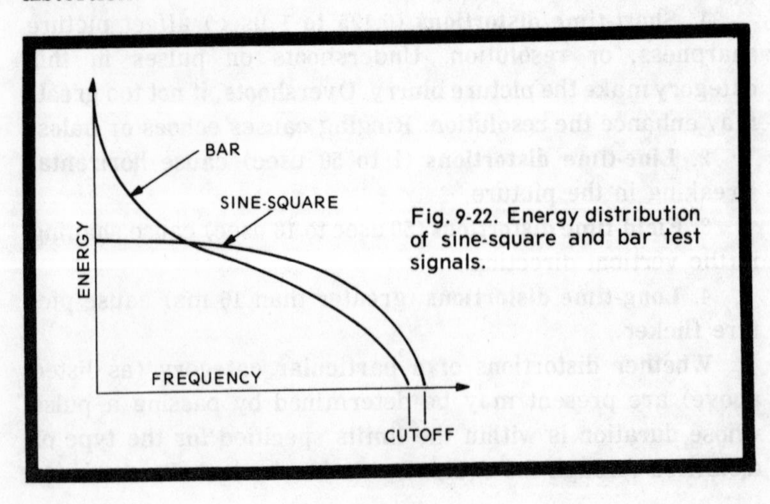

Fig. 9-22. Energy distribution of sine-square and bar test signals.

Another common TV test signal is the **sine-square bar**. This is actually an integrated sine-squared pulse, and in the time domain looks much like an ordinary square wave. Its energy distribution is different, however, as shown in Fig. 9-22. This test signal is usually derived by applying a very fast square wave to a sine-square shaping filter. The bar is used to measure line-time distortion. To facilitate these measurements using a waveform monitor, a special graticule is often used, which makes it possible to measure the waveform distortion in terms of a picture impairment factor, or K factor. The K factor may be considered a kind of "figure of **demerit**" for a television system.

When TV equipment is taken out of service for maintenance, the usual test signal for both field-time distortion and line-time distortion is the so-called "window." A window signal consists of a sine-square bar of about 25 usec duration, which is applied on about half the TV lines per field. Fig. 9-23 shows a tilt in such a signal (as viewed on a waveform monitor) due to 100 nsec RC coupling time constant.

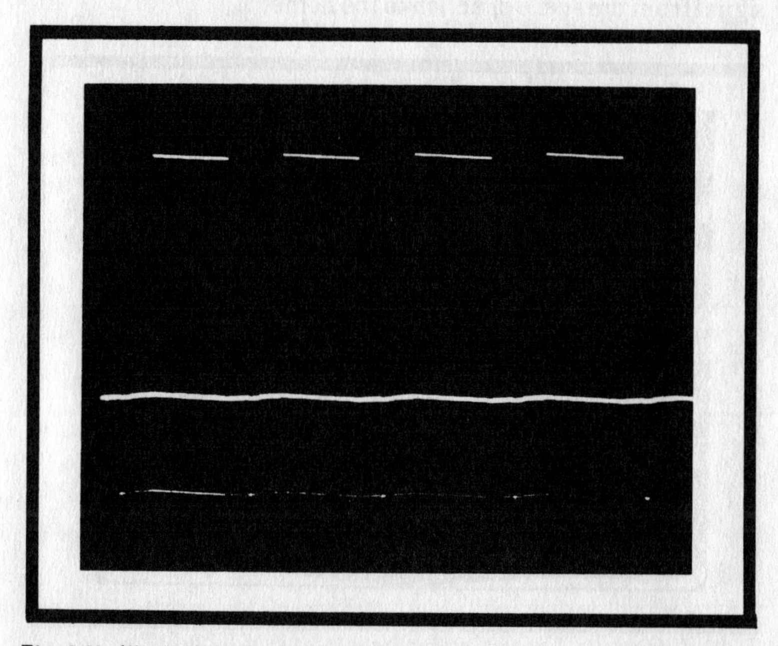

Fig. 9-23. Illustrating the tilt in the window signal due to a 100 msec RC coupling time constant.

Another television test pulse is the modulated sine-squared pulse of Fig. 9-24. This pulse measures two parameters important to color quality: **relative chrominance-to-luminance gain** and relative chrominance-to-luminance **delay**. Chrominance-to-luminance gain distortion causes errors in saturation, which are noticeable when viewing several stations or successive programs on one station. The home receiver can be readjusted to compensate for this distortion, but with the distortion kept low at the studio, the public does not have to correct it manually.

There is no adjustment on the home TV receiver to correct for chrominance-to-luminance delay, however. This is most easily detected with red lettering against a white background, with the red blurred and displaced to the right. This distortion can appear at almost any part of a TV system, but is especially important in TV transmitters and in CATV.

The measurements we have discussed are but of a few of the measurements made with TV waveform monitors to insure faithful reproduction and transmission of the video signal from the scene of action to the home.

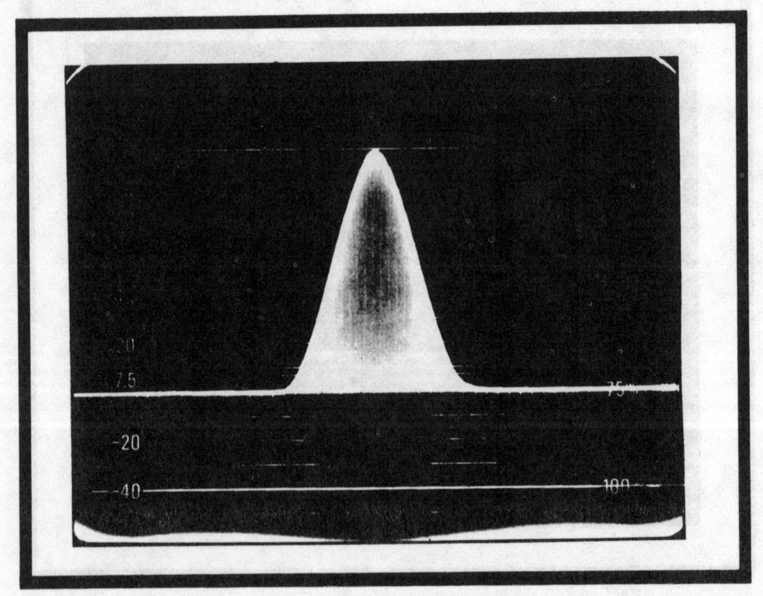

Fig. 9-24. The modulated sine-square pulse, displayed on a waveform monitor.

Sophisticated Techniques

Up to this point, we have considered the basic principles of scopes and of their operation, manufacturer specs, special features of scopes, and special-purpose scopes. In this chapter we will look at a few examples of sophisticated yet simple techniques of measurement using scopes.

STRAIN GAGE MEASUREMENTS

A strain gage is a measuring element used for converting force, pressure, tension, etc., into an electrical signal. A strain gage is basically a thin wire, which is attached to an object to which a force is applied so as to measure the effect of the force. As the strain-gage wire is stretched, its resistance will increase due to its increased length and decreased cross-sectional area. In Fig. 10-1 the fine wire, typically 1 mil

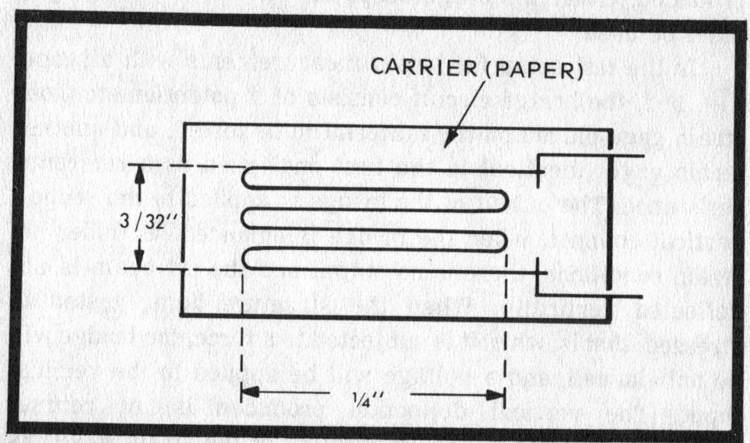

Fig. 10-1. A typical strain gage, consisting of a wire element cemented to a paper base.

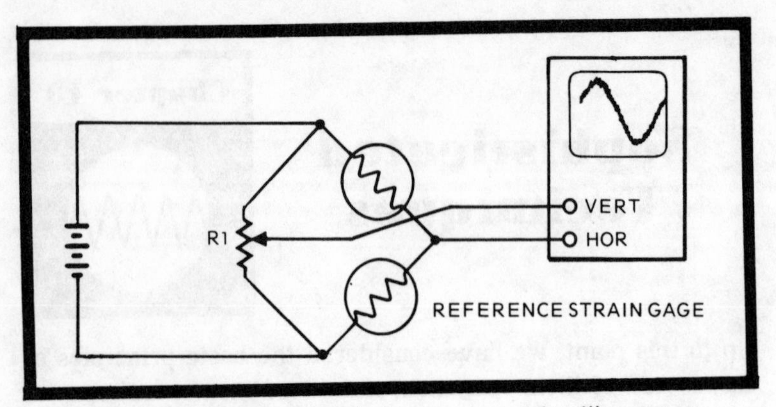

Fig. 10-2. A test setup for strain measurements with a scope.

diameter, is folded to form a grid, and encased in some base material such as paper or Bakelite. The paper or Bakelite base material is sometimes called the **carrier**. Gages come in grid lengths from under 1/16" to 6" or more, and grid widths from a single strand to 3/16" or so.

Strain gages are usually connected in a bridge circuit. When a static, or constant strain is to be measured, readout may be from a meter dial. Whenever the quantity being measured changes too fast to be measured with a meter, we have entered the realm of dynamic measurements, and we call for a scope. To record the instantaneous displays produced, either a storage scope or waveform photography must be used.

In the test setup for strain measurements with a scope, Fig. 10-2, the bridge circuit consists of a potentiometer, one strain gage placed on the material to be tested, and another strain gage (identical to the first one) used as a reference resistance. The output of the bridge is applied to the scope's vertical channel. When the bridge is balanced, as under no-strain conditions, there is no output and the crt beam is not deflected vertically. When the structure being tested is stressed, that is, when it is subjected to a force, the bridge will be unbalanced, and a voltage will be applied to the vertical input. The vertical deflection produced is, of course, proportional to the strain, or stretch, of the strain gage, vs time, as in Fig. 10-3.

Two terms that are very important in strain measurements and often misunderstood are **stress** and **strain.** Stress is force per unit area, or pressure, and is usually measured in pounds per square inch (psi). Strain is change in length per unit length, generally measured in microinches per inch.

In general, the following procedure may be used:

1. Build the test circuit of Fig. 10-2, placing the scope in operation. Use a sweep-time interval somewhat longer than the duration of the events to be observed. If a storage scope is not used, set up a scope camera to record the events.

2. Set the scope to measure dc voltage and adjust the potentiometer for zero deflection vertically, of the crt beam.

3. Stress (put force on) the structure under test, closing the camera shutter if a camera is used.

4. Using the storage tube trace or the developed photograph, measure the voltage vs time plot. With a calibration figure supplied by the strain gage manufacturer, the voltage may be converted to strain units, microinches per inch. The calibration figure is a certain number of microinches per inch per volt.

Strain gage techniques are used in measuring the stretching, bending, and twisting of materials; in measuring pressure, and in measuring acceleration. For measuring pressure or acceleration, strain-gage transducers are used in place of the strain gages shown in Fig. 10-2.

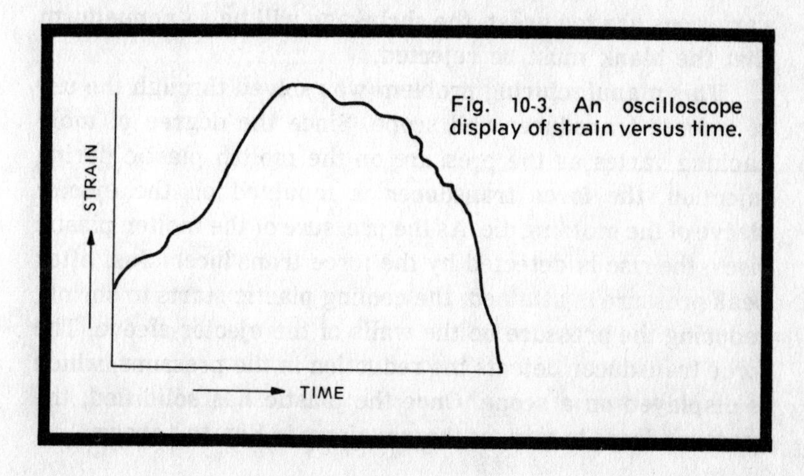

Fig. 10-3. An oscilloscope display of strain versus time.

PRODUCTION TESTING

The scope is increasingly used in electronics manufacturing plants for observing processes and making quality control tests of products. In conjunction with square-wave or sweep-signal generators, scopes are often used for checking reactive components such as transformers and chokes. Once assembled, electronic equipment is often sweep-tested for alignment, sensitivity, and other characteristics. Complex systems such as computers, transistorized controls, and radars are often tested using pulse generators and scopes. For testing equipment designed to operate at very high frequencies, sampling scopes are often used with electronic counters in production testing. The counters can give a digital indication of the rise time of pulses (in the equipment) lasting a fraction of a microsecond.

Manufacturing processes are often improved and maintained on a day-to-day basis by transducer measurement techniques using the scope as a readout device. One example encountered by the author is in the manufacturing of cam switches. In these switches there is a plastic drum having up to 40 individually operating cams. The drum is machined from an injection-molded blank, which must be annealed to insure the permanence of its dimensions. However, shrinkage results from the annealing process, and varies with the density of the molded part. Unfortunately, the variations cannot be detected until the annealing process is completed. If the density variations are too great, the shrinkage will be so nonuniform that the blank must be rejected.

This manufacturing problem was solved through the use of a force transducer and scope. Since the degree of mold packing varies as the pressure on the molten plastic during injection, the force transducer is mounted on the ejector sleeve of the molding die. As the pressure of the molten plastic rises, the rise is detected by the force transducer. Just after peak pressure is attained, the cooling plastic starts to shrink, reducing the pressure on the walls of the ejector sleeve. The force transducer detects the reduction in the pressure, which is displayed on a scope. Once the plastic has solidified, the pressure drops to zero, as the waveform in Fig. 10-4 shows.

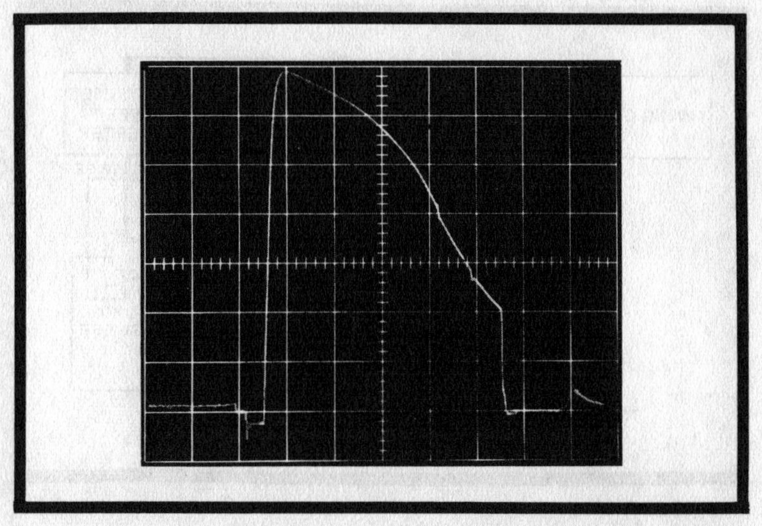

Fig. 10-4. Waveform produced by a force transducer during an injection molding process.

Experiments using the force transducer and scope soon showed the allowable tolerances for the waveform that would result in most economical use of the molding machine and yield parts of consistent quality. The reject ratio plummeted virtually to zero. The machine operator is now able to monitor each injection by monitoring the waveform and detect any problems immediately.

TIME-DOMAIN REFLECTOMETRY

Maintaining the fidelity of electronic signals during transmission is a primary concern of persons involved in designing, manufacturing, and maintaining electronic equipment. The coaxial cable is the most common means of transmitting signals from point to point. Time-domain reflectometry is a sophisticated yet simple means of checking the performance of coaxial transmission lines and locating faults in such lines.

A time-domain reflectometer is a radar-type device producing pulses and displaying reflections of the pulses. Fig. 10-5 shows a block diagram of a time-domain reflectometer. The pulse generator applies a pulse of one or two hundred volts

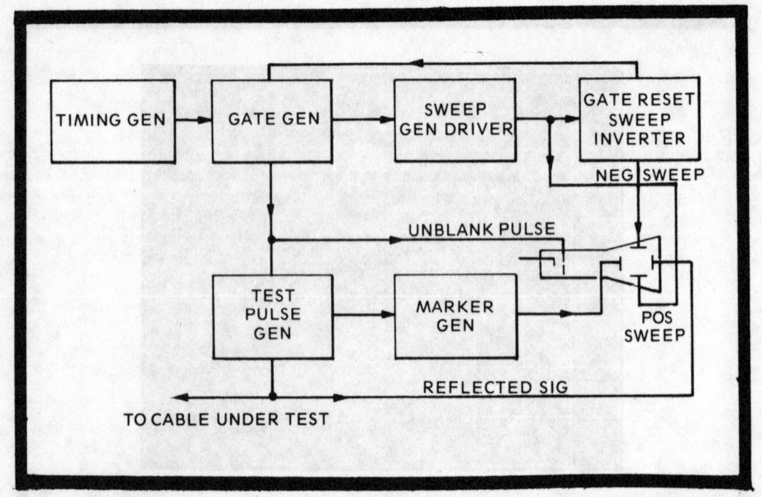

Fig. 10-5. A block diagram of a time-domain reflectometer, used for coaxial cable fault location and performance tests.

or so to the cable being tested. The pulse generator is coupled both to one end of the transmission line and to the vertical deflection plates of the crt, as shown. The terminating impedance for the other end of the line is selected so that the characteristic impedance of the transmission line is within its range.

When the terminating impedance is equal to the characteristic impedance, the waveform in Fig. 10-6A results. Only the original pulse is visible, since there are no reflections. Suppose the characteristic Z is unknown? All that is necessary is to connect the line as described above, and to vary the terminating impedance at the far end of the line until such a test pattern is observed.

In Fig. 10-6B, you see the type of pattern resulting when the terminating impedance is greater than the characteristic impedance. Fig. 10-6C shows the test pattern obtained when the terminating impedance at the other end of the line is less than the characteristic, or surge, impedance of the line. Note that in both cases the second pulse is much greater in amplitude than the first pulse. This is because the impedance at the input end of the line, where the pulse generator is connected, is quite a bit larger than the characteristic impedance of the line.

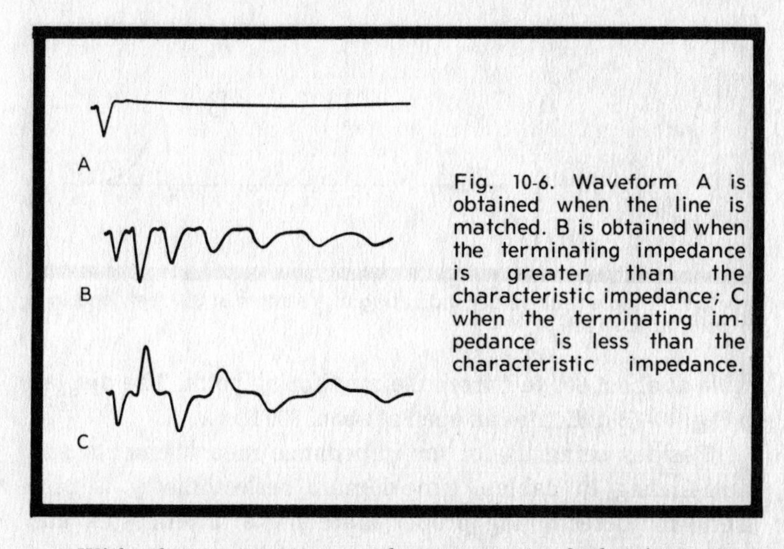

Fig. 10-6. Waveform A is obtained when the line is matched. B is obtained when the terminating impedance is greater than the characteristic impedance; C when the terminating impedance is less than the characteristic impedance.

With the test patterns shown, most of the important properties of transmission lines may be readily observed. Using a marker generator in the circuit of Fig. 10-5 permits the display of time markers on the crt. The transmission time along the line, that is, the time it takes a pulse to travel down the line, may be determined by measuring the distance between the first two pulses in terms of time markers. Once this time is known, the velocity of propagation may be found from the relationship $V = L/T$, where **L** is the length of the line, and **T** is the transmission time along the line **one way**. Knowing the velocity of propagation, the distance along the line to any discontinuity such as a short or open can be determined. When a discontinuity develops, it will show up as a previously unobserved pip on the waveform. The distance to the discontinuity may be found by solving for **L**. The type of fault may be determined by noting the polarity of the reflection. When an open exists, the pip representing the discontinuity will have the same polarity as the original pulse fed to the line. When a short exists, the pip will have a polarity opposite the polarity of the pip representing the original pulse.

In a cable-fault-finder, a graticule calibrated in feet greatly simplifies the location of a fault. Fig. 10-7 is an illustration of such a graticule with a maximum indication of 1200 feet. The display in Fig. 10-7A indicates a short in the

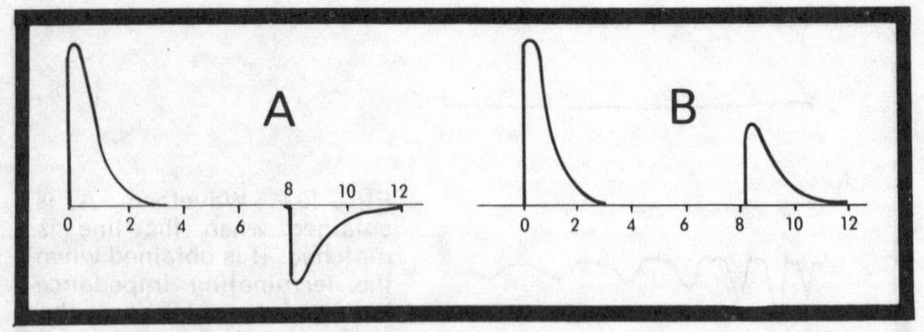

Fig. 10-7. A cable-fault-finder indicating in A a short at 800 feet, and in B an open at 800 feet.

cable at about 800 feet from the monitoring point. The display in Fig. 10-7B indicates an open at about 800 feet.

Besides being useful for impedance measurements and fault-finding in cables, time-domain reflectometry is also useful for determining proper matching of attenuators and junctions, lengths of coiled cable, relative attenuation of different cables, and whether intermittents exist in cable connectors.

The usefulness of time-domain reflectometry is greatly enhanced by screen photography, which permits a permanent record of transmission-line conditions to be made. A photograph of normal pulse conditions may be taken, and any new discontinuity may be detected by comparing the crt display with the photograph of normal conditions.

CRT SCREEN PHOTOGRAPHY

The most common form of waveform photography, the form we shall consider here, involves taking a still photograph of the crt screen. To eliminate ambient light from the photographs and to permit the waveforms to be photographed at close range without distortion, special scope cameras such as the one in Fig. 10-8 are used. To obviate the need for waiting for photographs to be developed, scope pictures are normally taken with Polaroid film, produced by the Polaroid Corp. Often, although not in the camera of Fig. 10-8, a means is provided for viewing the crt display during photography. The camera shown, while relatively inexpensive, takes excellent waveform photos and can be used with almost any scope.

One important consideration in waveform photography is exposure setting, which is an attempt to get just the right amount of light on the film to record the crt pattern. Even though the human eye and the camera differ greatly in their responses to light, it is normal practice to use the eye to determine the proper exposure setting. When the eye sees a continuous trace with no flicker or discernible spot movement, it is possible to make a good exposure estimate. In estimating exposure, first adjust the scope for normal brightness, that which seems proper for normal visual observation. Then set the camera f stop at 11 and exposure at 0.1 sec and take the picture. The use of the relatively small f stop suggested will give a good depth of field and should insure that everything is in focus on any scope. The relatively long exposure will insure that the entire sweep has time to "register" on the film.

Another important factor is the proper location of the starting point of the waveform by correct triggering. The object here is to pin down a specific, brief time interval out of infinity.

When trying to photograph a single, elusive transient, it is usually best to avoid using internal triggering, since noise or

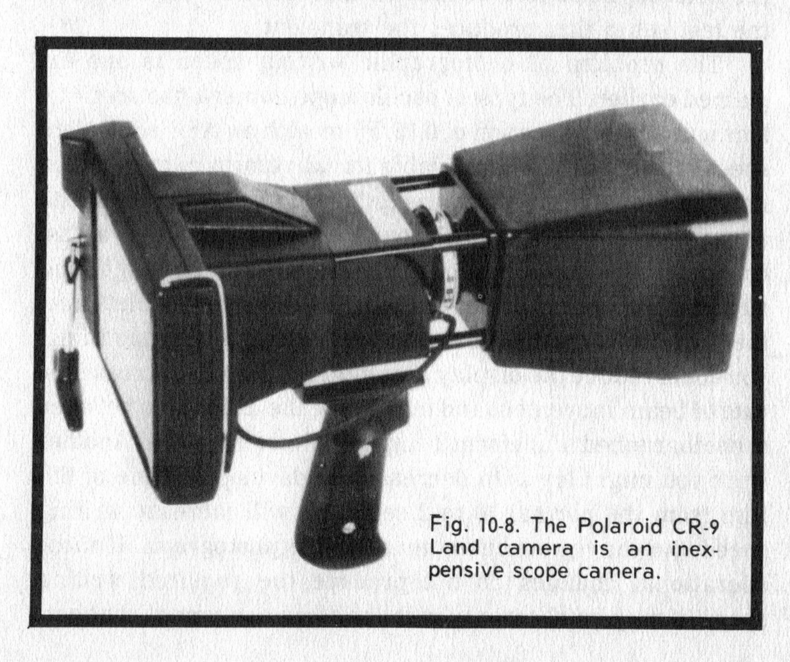

Fig. 10-8. The Polaroid CR-9 Land camera is an inexpensive scope camera.

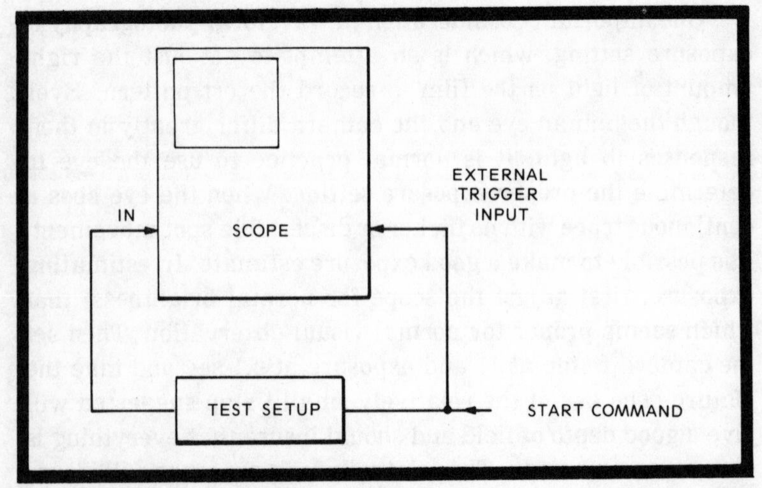

Fig. 10-9. The start command triggers the scope and causes the test setup to produce a transient in this hookup.

other spurious signals will often cause a false trigger. Usually it is best to externally trigger the scope with the same switch or circuit that initiates the transient, in the manner of Fig. 10-9. In Fig. 10-9, the scope is externally triggered by the same predictable, noise-free command that initiates the action in the test setup that produces the transient.

The problem of photographic writing speed is one explained earlier. The typical oscilloscope camera has an f/1.9 lens with a magnification of 0.85. Film with an ASA equivalent speed of 3000 is readily available for waveform cameras, and its high sensitivity permits recording high-speed phenomena as well as slow pulses and repetitive signals. When a fast transient is on the fringe of a camera's usability range, and when maximum crt intensity does not produce a usable trace, there are a number of measures you might try. For one thing, you could reduce the display amplitude. This will decrease the rate of beam movement and may make the difference between a photographed waveform being legible or illegible. Another trick you might try is to decrease the developing time of the film from the normal 10 to 2 sec. This will increase writing speed, but may produce an unattractive photograph. If these operational changes do not produce the required writing speed, it may be necessary to make some equipment changes.

Most scopes nowadays use a P31 phosphor. P31 phosphor is a very efficient emitter of light peaked at about 530 nanometers, an ideal wavelength for ordinary visual observation. For this reason, phosphor comparison charts such as the one way back in Table 2-1 often use P31 as a 100 percent reference for relative luminance or brightness.

Luminance values are brightness values measured through a C.I.E. Standard Eye Filter having the response characteristics of the human eye. Visual response, however, is not the same as film response. Although P31 is optimum for visual observation, it is not optimum for photographic writing speed. Also, the persistence, or after-glow, of P31, while normally an advantage, will often require a short wait to avoid photographing previous traces. Colored filters will reduce the persistence effects, but will also attenuate the total light output of the writing beam. A phosphor optimized for photography is P11. The light output of the P11 phosphor peaks (Fig. 10-10) at around 450 nanometers. Although this gives an optimum photographic writing speed, P11 has only 25 percent of the luminance of P31, and is far from optimum for visual observation. P11 has an additional photographic advantage of

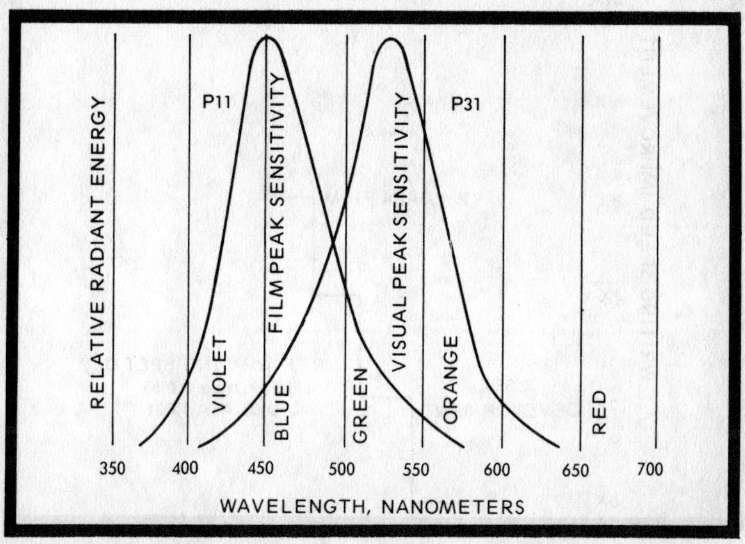

Fig. 10-10. The wavelength of the peak light is the key to phosphor selection. P11 is used mainly for waveform photography.

very short persistence. As shown in Fig. 10-11, type P11 phosphor gives a writing speed improvement of two times.

Writing speed may be improved by a factor of 2 or 2.5 by substituting 10,000 ASA film for the standard 3000 ASA film. One film of this type is Polaroid type 410 film, a roll film that is not usable with flat-pack backs.

If the general-purpose camera with its f/1.9,mag 1:0.85 lens is not fast enough, a faster f/1.2,mag 1:0.5 lens may be used, giving a writing speed improvement of perhaps four times. Faster scope cameras have a larger lens and usually a smaller magnification. As a result, light-gathering ability is more than doubled, and the light is concentrated on a smaller area of film by the smaller magnification.

A technique known as **fogging** is sometimes used to increase writing speed, but it is somewhat less convenient than the other methods given here. This is a sensitizing process used before, during, or after trace exposure; and consists of a

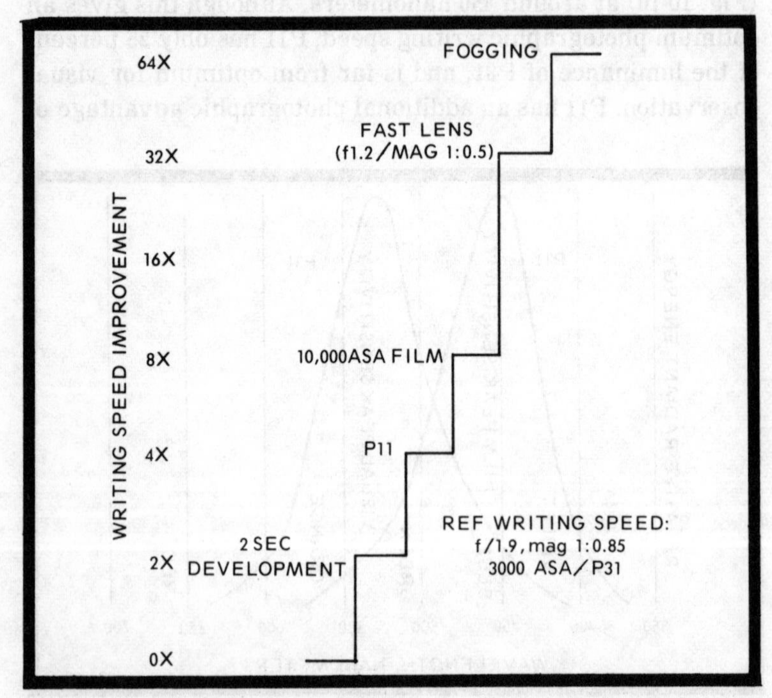

Fig. 10-11. The approximate writing speed improvement obtained with various practical techniques.

slight overall exposure of the film. While somewhat inconvenient, this technique can about double the photographic writing speed when compared to a fast lens.

CONCLUSION

As we have seen, there are many types of scopes, with many ways they can be used. We believe we have satisfied your needs for knowledge of how most scopes operate and how they are used. For more information about a particular scope or particular, perhaps unique, applications, query your electronics parts distributor or the scope manufacturer directly. You'll find that they will be very helpful.

Scope Applications

(Reprinted from TM 11-671)

A-m Transmitter Testing

Test oscilloscopes are used with transmitters for modulation monitoring, alinement, neutralization of r-f amplifiers, and general trouble-shooting. For modulation monitoring, the oscilloscope presents the instantaneous changes in modulation percentage.

a. Typical Test Set-Up. For most a-m transmitter tests, the r-f output of the transmitter (modulated or unmodulated) is fed directly to the vertical deflection plates of the oscilloscope. For alinement, neutralization, and similar operations, the output is fed to the vertical amplifier. A-m transmitters have high operating voltages in the r-f and modulator systems. The leads from the oscilloscope to these sections must be well insulated. The r-f voltages should be inductively coupled to the oscilloscope. Connections should be insulated against flashover to ground.

b. Block Pattern. The block pattern is one of the three types of pattern displayed on the oscilloscope for the output of an a-m transmitter. This pattern (A, fig. 199) is produced when the carrier is fed to the vertical-deflection plates of the oscilloscope with the internal sawtooth sweep of the oscilloscope set to a very low frequency. The modulating frequency is much greater than the sweep frequency. Consequently, the number of cycles of modulation displayed is too great to be observed individually. However, the presence of modulation is visible as horizontal light streaks across the rectangular block of light. This is shown in A, where two bright lines traverse the block pattern and divide it into three equal areas. This indicates 50 percent sine-wave modulation. One-hundred percent sine-wave modulation results in one light streak through the center of the block. These streaks result from the compression of many downward modulation peaks. Complex-wave modulation produces many more light streaks as in D. An unmodulated carrier has no light streaks.

Figure 199. Three types of pattern used for transmitter modulation testing.

c. Modulated-Wave Pattern. The second type of pattern presents the modulated-wave envelope. It shows the changes in carrier amplitude with time (B and E, fig. 199). This display is produced by feeding the carrier signal to the vertical-deflection plates of the oscilloscope and adjusting the internal time-base sweep to some submultiple of the modulation frequency. In B, the sweep frequency is one-fourth the modulation frequency. This trace shows 100 percent sine-wave modulation of the carrier. If the modulation percentage were less, the downward modulation peaks would be farther apart. The percent modulation is obtained from the following formula:

$$\text{Percentage} = \frac{B-A}{B+A} \times 100$$

where B is vertical height of the pattern at its widest point, and A is the vertical height of the pattern at its narrowest point.

d. Trapezoidal Pattern. The third type of pattern is shown in C and F, figure 199.

(1) This display presents changes in r-f carrier amplitude vertically against changes in modulating amplitude horizontally. The r-f voltage is applied to the vertical-deflection plates and the modulating voltage is applied to the horizontal-deflection amplifier input. To obtain the modulating voltage, connect a .1-Uf coupling capacitor to the modulation connection on the r-f amplifier. Also connect a voltage-divider resistor (about 5,000 ohms) between the capacitor and ground. Adjust the tap on the voltage divider to tap off a few volts of the modulator output voltage. The sweep-frequency selector of the oscilloscope is set to the OFF position. The percentage modulation formula for this pattern is the same as for modulated-wave patterns.

(2) The graphical development of the trapezoidal pattern is shown in figure 200. One cycle of the modulated-wave envelope combines with 1 cycle of the modulation voltage to form the pattern. The electron beam moves from left to right across the screen as a result of the change in voltage from 0 to 6 on the modulation

(horizontal deflection) signal. During the same time, the beam sweeps across the screen vertically many times because of the rapid changes in vertical-deflection voltage. These changes are a result of the changes in amplitude of the modulated r-f voltage. On the vertical-deflection wave, the points 2′ on the positive half and 2 on the negative half do not occur simultaneously. During the small time difference between them, the voltage on the vertical-deflection plates changes from the value which positioned the beam at 2 to the value of the modulated wave at 2′. This causes the beam to move upward and slightly to the right on the face of the tube to point 2′. The actual time difference between 2 and 2′ depends on the frequency of the r-f voltage fed to the vertical channel. In most cases it is very high, and the time difference between two corresponding points on the wave is small, making the line between them seem to be vertical. Each of the other lines of the trapezoidal trace is formed in the same way.

(3) After the beam reaches the right side of the screen, it moves from right to left because of the change in horizontal-deflection voltage from 6 to 12 on the modulating wave. Corresponding to these new horizontal-deflection voltages, the voltages on the vertical-deflection plates change according to the voltages 6′ to 12′ and 6 to 12 on the modulated r-f wave. For position 9 on the horizontal deflecting wave, for example, the line traced is 9 to 9′ on the trapezoid. The amplitude of the line is the same as the amplitude of the line from 3′ to 3, and it falls in the same place on the screen because of the symmetry of the modulating wave. For the pattern to have straight sides, the amplitude of the carrier must change linearly with the modulating voltage. Differences in phase between the modulator output and the modulated carrier result in the type of trace shown in F, figure 199.

e. Correct and Defective Operation. Proper transmitter operation results in patterns similar to those shown in A, B, and C, figure 199. If the amplitude-modulation percentage is less than 100 percent, the modulated-wave pattern shows shallower troughs than those in B. The trapezoidal pattern resembles the final trace shown in figure 200. Trapezoidal patterns generally are more useful for tracing defects than other types of patterns. Some of the causes for improper operation are too high or too low plate voltage in the final r-f stage, insufficient r-f excitation on the final r-f stage, mismatch between modulator and r-f final stage, imperfect neutralization, and distorted modulation voltage, shown in D. These produce distorted patterns which can be correlated with the transmitter defect.

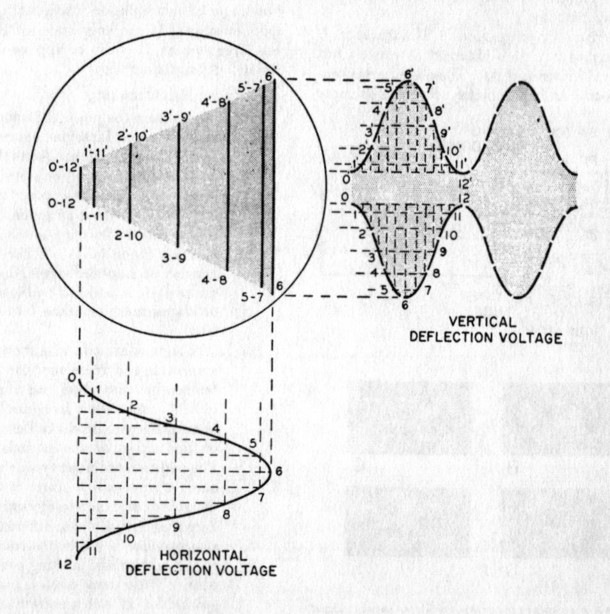

Figure 200. Graphical development of trapezoidal pattern.

Vibrator Testing

The vibrator is used in electronic equipment to change d-c to a-c. Basically, it is a high-speed reversing switch. When one set of contacts is closed, the voltage output is in one direction; when the second set of contacts is closed and the first set is open, the voltage output is in the opposite direction. The time during which a set of contacts is closed is called closure time. The vibrator output is fed to the primary of a step-up transformer. The stepped-up voltage across the secondary is rectified and filtered and serves as the plate-voltage power supply for the equipment of which it is a part. If the vibrator itself rectifies the output of the transformer it is called a synchronous vibrator. If a rectifier tube is used to rectify the transformer output, the vibrator is a nonsynchronous one.

a. Nonsynchronous-Vibrator Power Supply. In A of figure 201 is the schematic diagram of a typical nonsynchronous-vibrator power supply. This type of power supply often is used in motor-vehicle communication equipment. In such cases, the storage battery of the vehicle furnishes the low-voltage d-c supply to the vibrator. The stationary vibrator contacts are 1 and 2. The vibrating contact is 3. Faulty vibrators can result in lowered B-supply output voltage, increased battery drain, arcing, and noise. For testing a nonsynchronous vibrator, use a 5,000-ohm load resistor to simulate normal operating conditions. When testing vibrators with an oscilloscope, the internal sync and sweep voltages are used. The input is fed to the vertical amplifier. The sweep is set to some submultiple of the vibrator frequency. The test leads should be shielded to prevent stray pick-up.

b. Patterns of Operation. The traces in B and E, figure 201 were obtained at various test points in the circuit of A. These are normal operating patterns for the nonsynchronous-vibrator power supply. B shows the voltage taken from P1 to P3 of the primary of the step-up transformer. The voltage across P2–P3 is shown in C. The vibrator producing these patterns had a relatively high closure time. The sum of the time durations of both the positive and negative plateaus is almost the total time for the whole cycle, showing that the contacts are close together. If the contacts are far apart, the vertical lines slope more gradually. D and E show the voltage output across S1–S3 and S2 and S3 of the secondary of the step-up transformer.

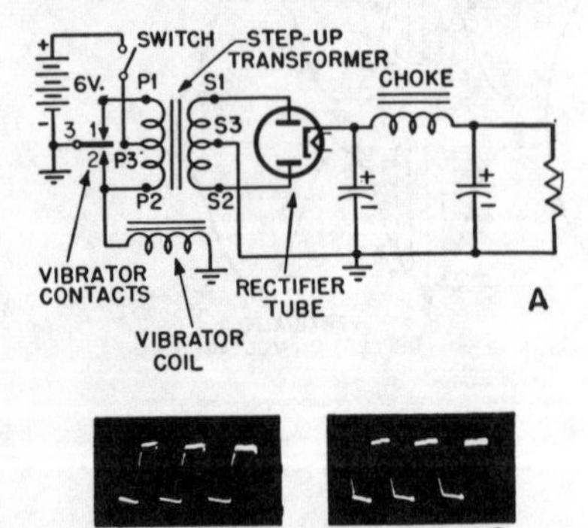

Figure 201. Nonsynchronous vibrator power supply circuit and typical operating patterns.

Lissajous Figures

a. General. The normal input to the horizontal channel of the oscilloscope is a sawtooth voltage which has a linear relation with respect to time. The sawtooth sweep, however, cannot be used for the precise determination of the frequency and the phase of a signal. For this application of the oscilloscope, a standard signal of power-line frequency or the output of a calibrated signal generator is applied to the horizontal channel of the oscilloscope. The signal to be observed is applied to the vertical channel. If both these signals are a-c voltages, the resulting pattern is called a *Lissajous figure*. The circular J-scan produced by two sine waves figure 109 is an example of a Lissajous figure.

b. Signals on Both Plates. Lissajous figures present the frequency or the phase relationship (or both) between the two a-c signals forming them. The amplitudes of the two signals applied to the oscilloscope are made equal in order to produce a pattern based on frequency and phase relationships. If it is not possible to have the amplitudes equal, the amplitude relationship should be held constant. If the amplitudes of the signals are large enough, they can be applied directly to the deflection plates.

c. Phase Measurement.

(1) To measure the phase difference between two signals, a Lissajous figure is set up. If the input signals are fed to the vertical and horizontal deflection amplifiers, these must both have the same number of stages. The output of an amplifier stage is opposite in polarity to the input. If the two channels do not have the same number of amplifier stages, the apparent phase shift introduced by the extra stage or stages must be taken into consideration.

(2) If the signals are two sine waves of equal amplitude and frequency, the phase difference between them can vary from 0 to 360°. As shown in figure 109, a 90° phase difference produces a circular trace on the screen of the cathode-ray tube. The addition of these two sine waves produces a trace which starts at the top of the screen and moves clockwise to produce a circle. A phase difference of 270° also produces a circle, the trace starting at the bottom and moving counterclockwise. These two circles are indistinguishable. If such a pattern is produced in comparing two signals on the oscillo-

scope, the phase difference is taken to be either 90° or 270°.

(3) A of figure 202 shows the patterns resulting from the addition of two sine waves with phase differences from 0 to 360° in steps of 22.5°. Sine waves of equal amplitude and frequency produce straight line images for phase differences of 0°, 180°, and 360°. Phase differences between 0–90°, 90–180°, 180–270°, and 270–360° produce ellipses of continuously varying characteristics, illustrated in A. Phase differences other than those shown produce similar ellipses.

(4) For more precise calculation of the phase difference between two sine waves of equal amplitude and frequency, two measurements must be made on the Lissajous figure. These consist of measuring the two distances a and b indicated in B, figure 202. For the ellipse shown, a is equal to seven spaces along the X axis and b is equal to eight spaces along the X axis; a divided by b gives the sine of the phase angle. In this example, the sine is seven-eighths, or .87, and the phase angle is approximately 60° or 300°.

(5) Phase differences between signals other than sine waves can also be measured. Two triangular waves result in a slanted line, a slanted rectangle, or a square standing on one of its corners, for phase differences of 0°, between 0° and 90°, and 90°, respectively. The same patterns are produced in reverse order from 90° to 180° phase differences. The same order is followed for the phase differences between 360° and 180°. Sine waves can be compared to triangular waves and square waves. Square waves can be compared with the other two types.

d. Frequency Measurement. Two input signals of different frequency produce a characteristic Lissajous pattern on the screen. These patterns are used to determine the frequency of one of the signals when the frequency of the other is known. The signal of unknown frequency is applied to the vertical amplifier of the oscilloscope. Signals from a generator capable of precise frequency settings are applied to the horizontal amplifier. If the two signals are sine waves, the pattern produced is a Lissajous figure similar to those shown in A, figure 203. If the two signals have the same phase and amplitude and if their frequencies are the same (that is, if the ratio of their frequencies is 1:1), the result is a circle on the screen. Frequency ratios are expressed in terms of vertical (unknown) frequency to horizontal (standard) frequency. If the unknown signal has a frequency twice that of the signal generator, the ratio of the frequencies is 2:1; the resulting pattern is shown in A, figure 203. To obtain the ratio from the pattern, count the number of horizontal tangent points and divide it by the number of vertical tangent points. If the unknown signal frequency is three times the signal generator frequency, the pattern will appear as one of those in B, figure 203. The pattern observed depends on the phase relationship between the two signals. In the closed pattern, the frequency relation is determined by counting the loops touching the horizontal tangents and dividing this by the number of loops touching the vertical tangents. To obtain the 3:1 ratio for the open pattern in B, divide 1½ (the open end of the pattern is called a one-half loop) by ½. This gives 3:1. To obtain the frequency of an unknown signal which forms a 3:1 Lissajous pattern with a frequency standard of 2,000 cps, multiply the standard frequency by the ratio—in this case, 2,000 times 3. The result is the unknown frequency. In this example it is 6,000 cps. The same method is used with other ratios.

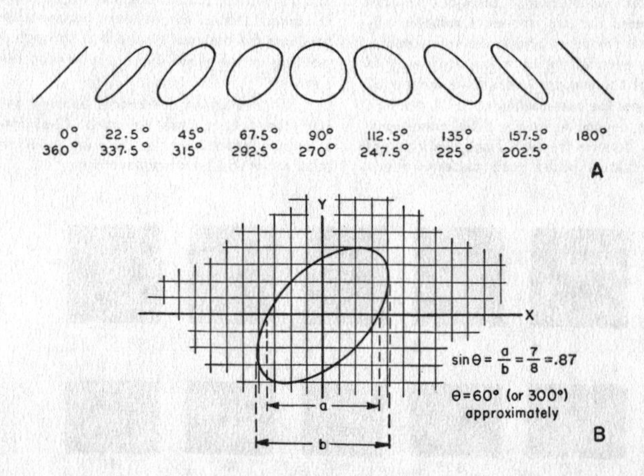

A, Different resultant patterns when the phase difference between the unknown and standard signals varies in steps of 22.5; B, Method of calculating the phase difference of two input signals from the pattern.

Figure 202.

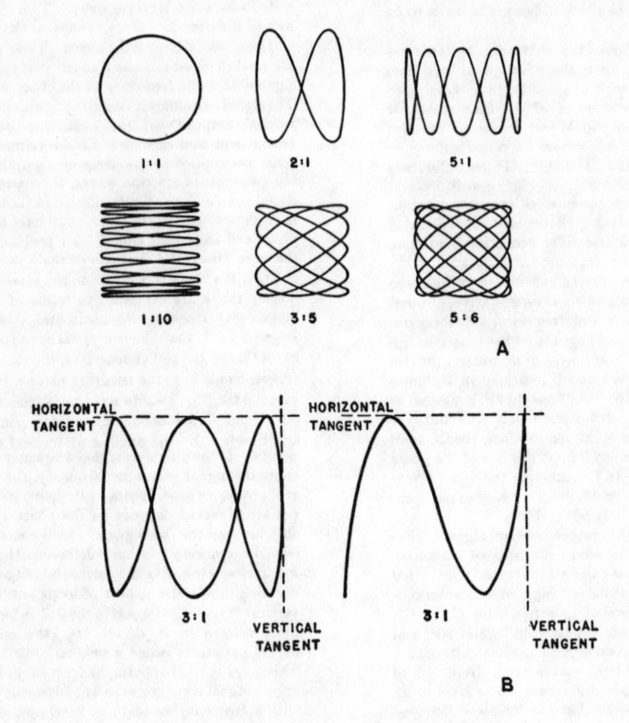

Figure 203. Lissajous patterns for various frequency ratios.

e. COMBINED OBSERVATIONS. Lissajous patterns can be produced for two sine-wave voltages differing in both frequency and phase relationship. A sine-wave vertical signal whose frequency is twice that of the standard sine-wave horizontal signal produces the patterns shown in A through E, figure 204, depending on the phase conditions. The phase difference for each Lissajous figure is shown immediately under each pattern. Similarly, a vertical unknown signal whose frequency is one-half that of the standard horizontal signal produces the patterns shown in F through J, depending on the phase difference between the two signals.

For phase differences between 180° to 360° the patterns are reversed. Characteristic Lissajous patterns such as these are produced for other ratios and for other waveforms.

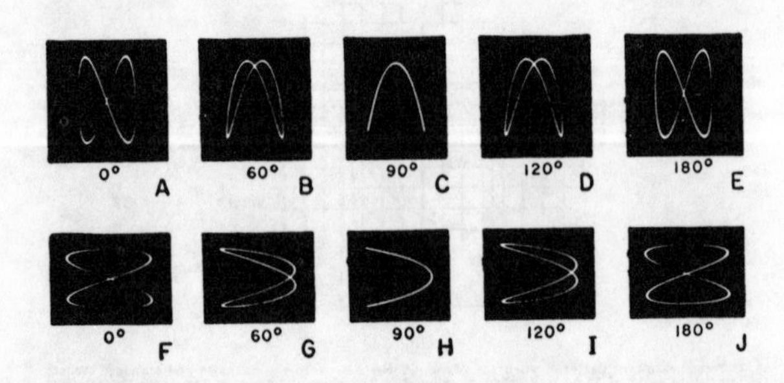

Figure 204. Lissajous patterns for two signals varying in frequency and phase.

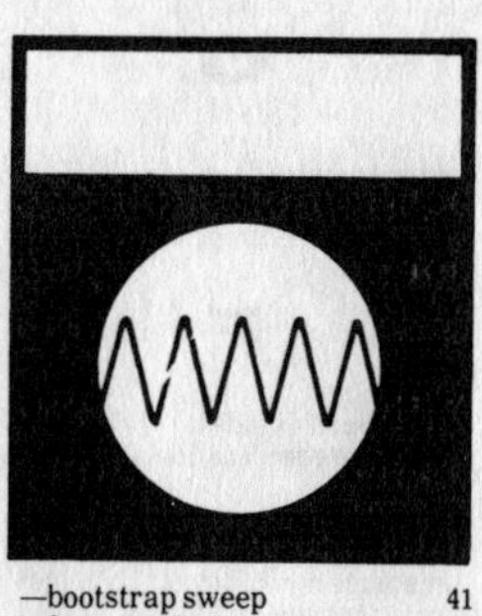